Library of
Davidson College

Wealth Redistribution and the Income Tax
Principal Paper by Norman B. Ture

A Liberty Fund Seminar
Administered by
The Law and Economics Center

Wealth Redistribution and the Income Tax

Principal Paper by Norman B. Ture

Edited by
Arleen A. Leibowitz

Lexington Books
D.C. Heath and Company
Lexington, Massachusetts
Toronto

Wealth Redistribution and the Income Tax presents papers from a seminar held at the Law and Economics Center in January, 1977 as part of the program of Liberty Fund, Inc. Liberty Fund is a foundation established to encourage study of the idea of a society of free and responsible individuals.

Library of Congress Cataloging in Publication Data

Main entry under title:

Wealth redistribution and the income tax.

    Papers and discussions from a conference sponsored by the Liberty Fund, Jan. 21-23, 1977.
    Bibliography: p.
    1. Income tax—United States—Congresses. 2. Income distribution—United States—Congresses. 3. Wealth—United States—Congresses. I. Leibowitz, Arleen A. II. Liberty Fund.
HJ4652.W54   336.2'42'0973   77-18652
ISBN 0-669-01506-7

Copyright © 1978 by D.C. Heath and Company.

All rights reserved. No part of this publication may be reproduced or transmitted in any form or by any means, electronic or mechanical, including photocopy, recording, or any information storage or retrieval system, without permission in writing from the publisher.

Published simultaneously in Canada.

Printed in the United States of America.

International Standard Book Number: 0-669-01506-7

Library of Congress Catalog Card Number: 77-18652

Contents

	Preface	vii
	Principal Paper	1
Chapter 1	Taxation and the Distribution of Income Norman B. Ture	3
	Commentaries	43
Chapter 2	Commentary: *Michael J. Graetz*	45
Chapter 3	Commentary: *Martin Feldstein*	57
Chapter 4	Commentary: *Boris I. Bittker*	65
Chapter 5	Commentary: *Richard Musgrave*	75
	Discussion	85
	Philosophical-Ethical Issues	87
	Bias against Capital Formation	93
	Capital Formation and Growth	99
	Redistribution	105
	Consumption Tax versus Income Tax	115
	Tax Rates	123
	Conclusion	127
	Bibliography	129
	About the Contributors	131
	About the Editor	133

Preface

There is a growing consensus that the time has come for a complete rethinking of our tax system. This rethinking would involve not merely an alteration in rates or deductions, but a restructuring of the base on which we levy taxes. What is called for is quite different from the Tax Reform Acts of 1969 and 1976, which modified but did not fundamentally change the existing tax structure. Such a change has been detailed in *Blueprints for Basic Tax Reform,* published by the United States Treasury in January 1977.

If the tax structure is to undergo an essential alteration, both the tax base and the philosophical basis for redistributive taxation should be scrutinized. Recognizing that a reformulation of the tax system should include a reevaluation of the relationship between the tax system and redistributive government policies, the Liberty Fund sponsored a conference January 21-23, 1977, at which lawyers and economists came together to address these issues. The conference was administered by the Law and Economics Center of the University of Miami School of Law. The five papers in this volume were delivered at that conference, and the dialogue that follows was distilled from the recorded interchange of ideas among the participants at the conference.

In the central paper in this volume Norman Ture returns to fundamentals. He reevaluates the ethical bases of redistributive taxation and concludes that our current system discourages the acquisition of capital. To alleviate this bias against saving, Ture argues in favor of taking consumption rather than income as the tax base. Commentaries on Ture's paper by Michael Graetz, Martin Feldstein, Boris Bittker, and Richard Musgrave modify and dispute some of Ture's points, but support his argument that fundamental changes in the tax system are warranted.

The prepared papers sparked a wide-ranging dialogue, an edited version of which follows the papers. In addition to debates on Ture's analysis of the ethical basis of redistribution and the effect of taxation on savings and investment, the discussions extended consideration to the impact of taxation on growth, the proper role of wealth in the consumption tax base, and the optimal rate structure. The conference participants were

William D. Andrews, Professor of Law
Harvard University

Martin J. Bailey, Professor of Economics
University of Maryland

Boris I. Bittker, Professor of Law
Yale University

Geoffrey Brennan, Visiting Professor of Economics
Virginia Polytechnic Institute and State University

Oswald Brownlee, Professor of Economics
University of Minnesota

James M. Buchanan, Professor of Economics
Virginia Polytechnic Institute and State University

Charles T. Clotfelter, Professor of Economics
University of Maryland

George Cooper, Professor of Law
Columbia University

Robert Eisner, Professor of Economics
Northwestern University

Alan L. Feld, Professor of Law
Boston University

Martin S. Feldstein, Professor of Economics
Harvard University

Charles O. Galvin, Professor of Law
Southern Methodist University

John T. Gaubatz, Professor of Law
University of Miami

Charles J. Goetz, Professor of Law
University of Virginia

Michael J. Graetz, Professor of Law
University of Virginia

Daniel I. Halperin, Professor of Law
University of Pennsylvania

Robert Hellawell, Professor of Law
Columbia University

William A. Klein, Professor of Law
University of California, Los Angeles

Charles E. McLure, Jr., Professor of Economics
Rice University

John K. McNulty, Professor of Law
University of California, Berkeley

David L. Meiselman, Professor of Economics
Virginia Polytechnic Institute and State University

Peter Mieszkowski, Professor of Economics
University of Houston

Richard Musgrave, Professor of Economics
Harvard University

Daniel Orr, Professor of Economics
University of California, San Diego

Harvey S. Rosen, Professor of Economics
Princeton University

Emil M. Sunley, Jr.
The Brookings Institution

Stanley Surrey, Professor of Law
Harvard University

G.S. Tolley, Professor of Economics
University of Chicago

Norman B. Ture
Norman B. Ture, Inc., Washington, D. C.

Alvin C. Warren, Jr., Professor of Law
University of Pennsylvania

Bernard Wolfman, Professor of Law
Harvard University

Thanks are due to all those who attended the conference for sharing their expertise and opinions. This volume has been published so that the informed views of lawyers and economists, confronting together the problems of taxation and redistribution, can be brought to bear in the upcoming debate on tax policy.

Arleen A. Leibowitz
Law and Economics Center
Miami, Florida

Principal Paper by Norman B. Ture

1 Taxation and the Distribution of Income

Norman B. Ture

The Issues: Ethics and Economics

Toilers in the fields of the soft sciences have a regrettable proclivity for accepting premises without critical examination. All of us concerned with tax theory and policy rely heavily on "everybody knows" propositions, often without being conscious of the fact. When we go out of our way to identify and to analyze rigorously these basic assumptions, challenging and exciting conclusions often emerge. And even when we come away from such exercises without having reached solid conclusions, the questions that we have raised are themselves fascinating. For this reason we owe much to Blum and Kalven for the questions that are the outcome of their study *The Uneasy Case for Progressive Taxation*. The challenges they posed in that searching examination of entrenched doctrine have not yet been met head on—either in the scholarly literature or in the public policy forum. I take this neglect as evidence of the capacity of ideology to absorb or deflect the most punishing blows. Income redistributive public policy, particularly tax policy, has withstood the assaults of abstract and empirical analysis demonstrating its failure; whether aimed at economic stabilization or at minimizing avoidance, tax legislation manifestly remains geared to some unspecified vertical equity criterion.

Ideologies need periodic challenges to avoid fatigue. This seminar, we hope, will refresh the debate about the appropriateness and feasibility of income redistribution as a goal of tax policy. The 1976 tax reform legislation was a devastating reminder of the power of egalitarianism in shaping public policy. It is time to take out the "everybody knows" notions on which the egalitarian prescriptions for tax policy depend and to seek to determine whether they are indeed so valid as basic premises for the arguments of tax policy that they may be vouchsafed forevermore.

The issues concerning any such discussion fall under two familiar headings—ethical-aesthetic and economic—although the distinction between the two sets of issues has been blurred by the fact that economists have treated the first much as if it was a question of economic analysis. This is not to say that one who has chosen economics as a profession should be debarred from discussion of the ethical or aesthetic attributes of any policy question. Nor is it to say that there are no aspects of the question to which economic analysis may be appropriately applied. Rather, it is to say that skill in economic analysis provides no comparative advantage for examining many of the questions of "goodness" or "badness" or the "beauty" or "ugliness" of a public policy. I shall try to confine my observations on this first set of issues to a few of the questions for which economics is a useful discipline.

Ethics and Aesthetics: What Do They Require of Taxes and the Shape of the Income Distribution?

The first set of concerns comprises the matters of ethics-aesthetics from which the equity criteria of tax policy ostensibly are rigorously derived. Income redistribution is one step removed from the immediate announced objective of redistributive tax policy, namely a fairer distribution of tax burdens. The tax policy debates seldom if ever explicitly cite reducing the Gini coefficient of a Lorenz curve of after-tax income as the objective of tax reform; the flights of oratory are instead launched at the tax privileges of affluent individuals and big corporate businesses, which allegedly permit them to carry less than their fair share of the total tax burden and hence require the rest of us to assume too large a share. Except when the effective tax rates of those identified as tax-privileged are shown to be less than those of the rest of us, however, "fair shares" of any tax total have to be the ultimate in imprecision. What, other than exact equality in effective rates, would be the standard of fairness of the distribution of tax burdens? Those in pursuit of a fairer distribution of tax burdens vociferously reject the strict proportionality of tax burden to income, wealth, or consumption as the appropriate standard. It is surely appropriate to infer, therefore, that behind the efforts to make the affluent pay more taxes—and presumably the nonaffluent to pay less—must lurk some notion of the desirability of greater equality in some relevant measure of economic status.

On what considerations do any such desiderata rest? For an extended period economists sought to apply the calculus of utility maximization to the determination of optimum schedules of tax liabilities, that is, to the determination of the optimum degree of income equalization. As formal exercises in logic (providing one ignored premises), these efforts were often highly satisfying aesthetically and ethically. In a more recent era, however, the premises were exposed to rigorous examination and were revealed to be far too shaky a foundation for the elaborate edifice of progression in tax liabilities.

Many of the challenges posed by modern economics to earlier theories of optimum tax burden distribution are now familiar to scholars in the field of tax and fiscal theory. The challenges, just as the views they oppose, warrant periodic reexamination. Not in this discussion, however; fascinating though they are as intellectual puzzles, even cursory attention to them would take me too far afield and unduly extend this paper.

There is, however, one seldom articulated issue of redistributive fiscal policy that bears directly on the central theme of this paper. The issue is best posed as a paradox. While equality of economic status may be a desideratum and feasible for an impoverished society, its attainment in such a society is of little consequence; but in an affluent society, where equalizing economic status would be significant, it is neither desirable nor feasible. To be sure this observation rests on hyperbolic characterizations of impoverished and affluent societies; exception should certainly be taken to it. But the thrust of the argument is little affected by more precise specification.

The difference between poverty and affluence is not solely a matter of production endowments per capita; it is also a matter of the diversification and specialization of demands and production agencies. Impoverished communities have relatively limited amounts of production inputs, but their wants and the meager production resources used to satisfy them are little differentiated. In poor societies the frequency distribution of characteristics denoting economic status is extremely skewed; there are very few affluent persons, and the Lorenz curves of such characteristics are not smooth arcs below the 45° diagonal but are more nearly horizontal over most of the distribution, rising sharply at its tail end. Those few affluent persons occupying the latter part of the distribution are particularly offensive since they so sharply contrast with everyone else and since their affluence very likely consists of relatively many of the few things that comprise the society's output and are deemed essential to support life.

In such a society there is no great trick to defining equality of economic status. By assumption the population is to a high degree homogeneous with respect to relevant economic attributes. Economic equality can be closely approximated in terms of the claims to the rudimentary products and services that comprise the economy's output. If equalization is the goal of public policy, it is neither difficult to identify nor to implement. On the other hand, neither is it significant. Locating everyone on the 45° diagonal means little, if anything, to anyone's well being, except that of the formerly affluent.

Affluent societies, by contrast, are highly diverse and highly specialized as well as richly endowed. The pursuit of economic equality, therefore, confronts enormous difficulties. The greater the diversity and specialization of products, services, and production inputs, the greater the heterogeneity of consumption patterns; and the greater the dispersion in effort-leisure and saving-consumption choices, the greater the conceptual problems in identifying appropriate criteria of equality of economic status.

I believe that this problem has not been addressed rigorously. We have long relied on the proposition that income is the best measure of ability to pay taxes; if this proposition is addressed to the use of taxes to redistribute income, it implies either that income, as defined ideally for tax purposes, is a good proxy for economic status or that the proper objective of redistributive policies is not to equalize status—the ownership of stocks of claims to resources and output—but to equalize the flows thereupon. For what purpose would we want to equalize these flows, income, in a free society? The implicit answer is to assure equality of consumption, since if this is not the case and if capital has a positive marginal value product, the distribution of income will immediately become unequal unless diverse individuals make identical saving-consumption choices from identical disposable incomes. If income-equalizing tax policy perfectly cancels differences in saving-consumption choices, no private saving results;[1] the ultimate goal of income equalization, wittingly or not, is in fact consumption equalization. If this is indeed the objective, how can it be realized without homogenization of the society?

One of the major elements of heterogeneity in our society is the choice between leisure and the explicit rewards for market-channeled effort. Barring private saving, an income-equalizing and hence consumption-equalizing fisc must override individual choices between consumption and leisure. That is, those who prefer relatively large amounts of leisure would nevertheless be endowed by the fisc with a consumption "capacity" equal to those who prefer relatively large amounts of consumption. If we ignore for the moment the fiscal effect on the individual's choice between consumption and leisure, consumption equalization necessarily means that those who prize consumption highly relative to leisure must consume less than they would prefer in order to increase the consumption of those who place a lower store on consumption relative to leisure. The "goodness" or "beauty" of these results eludes me. What is there about equality and sameness of consumption that is either beautiful or just in a society whose members, if allowed, seek to distinguish themselves in this and other respects?

On this line of reasoning, equalizing economic status in an affluent, diverse society is not only beset with the difficulties of identifying targets and implementation; it is also repugnant, ethically and aesthetically. If these obstacles were overcome, however, an equalizing fisc would have significant consequences, in contrast with the poor society in which redistribution might well be desired and feasible but inconsequential in effect.

Here is the very essence of the ethical-aesthetic issue of redistributive policies in an affluent, hence diversified, society. Diversity with respect to virtually any cultural or economic attribute is mirrored by diversity in the type as well as amount of property rights. If people are to be free to exercise their preferences—if their penchant for differentiation is not to be institutionally constrained—the values of differing property rights will change through time unless preferences are once and for all fixed. But changes in the relative values of different property rights preclude equality in economic status unless everyone is required, irrespective of preferences, to hold identical portfolios. In other words, achieving and preserving such equality requires either that the diverse portfolios of property rights be equally distributed, once and forevermore precluding differences in preferences or in efforts to effectuate them, or that differing portfolios each once and forevermore have the same aggregate value.

Clearly neither of these alternatives may be obtained in a free society that relies on a free market system for the evaluation and distribution of property rights. Equalization of economic status and economic freedom are therefore incompatible. An institutional thrust toward economic equality is necessarily equivalent to the elimination of the free-market mechanism for establishing values, organizing production subject to efficiency maximizing constraints, distributing the output of production, and providing for rising living standards.

Moreover, since differences in the type and amount of property rights of different individuals are functionally related to differences in their personal attributes, there is no way to prevent an institutional thrust toward equality of economic status from extending far beyond that highly elusive goal and to avert its constraining our penchant for differentiation in every aspect of life. We do not deem sameness of cultural attributes an ethical or aesthetic plus. We do not insist that each person should strive to find his identity exclusively in the characteristics common to us all. We do not hold that it is ugly or bad to seek distinction. If we seek economic equality, however, we are in effect asserting that the only acceptable distinctions are those that are deemed, contrary to fact, to have no relevance for economic status. Unless it can be demonstrated that public policies that succeed in equalizing us economically would not at the same time herd us into sameness in all other aspects of our culture, we should find such policies ethically and aesthetically distasteful.

I think we do. I think it is our instinctive apprehension that such policies would rob us of the power of effective discrimination and differentiation that accounts for the disguises worn by redistributive tax proposals. Such proposals will distribute taxes more "fairly," we are told. Negative income taxes will be more "efficient" than other types of welfare programs, it is asserted. Money earned by money should be taxed at least as much as money earned by labor, we are sloganeered. Abatements of excessively punitive taxes on savings are called "loopholes" and are characterized as available only to the rich and big business. When the subject is addressed to an audience comprising all of us, it is not represented to us as concerned with equalizing our incomes, our wealth, or our consumption. We must be deluded into accepting policies and institutional arrangements that frustrate our efforts to distinguish, to differentiate, to follow our own stars. Redistributive fiscal policies, undisguised and naked, may be things of beauty to elitists. For all others they must be clad in shimmering sequins that from a distance appear to be the stuff that halos are made of.

To recapitulate, I argue that the prescriptions for income redistribution in societies such as ours are not well conceived. Redistributive policies really aim to equalize consumption, an objective which, if explicitly identified, would be rejected out of hand.

Egalitarians conventionally deny this unlovely result of equalization. Arthur Okun, for example, asserts that "equal standards of living would not mean that people would choose to spend their incomes and allocate their wealth identically. Economic equality would not mean sameness or drabness or uniformity because people have vastly different tastes and preferences."[2] Okun's assertion is useful as an expression of the repugnance we all feel toward the "sameness," "drabness," or "uniformity" that he claims is not a necessary result of economic equality and which I believe is an inevitable consequence of maintaining the state of economic equality. Clearly Okun does not intend economic equality to interfere with people's exercise of their "vastly different tastes and preferences," provided that any such exercise does not result in differences in economic status. The obvious question is what ethical or aesthetic system drives to the conclusion that variety of preferences and of efforts to realize differing personal goals is good and beautiful, except for the preference and associated activities of distinguishing oneself economically? Okun, and he is not alone in this connection, supplies no answer to this question but merely states his preference: "Incomes that match productivity have no ethical appeal. Equality in the distribution of incomes . . . would be my ethical preference. . . . This preference is a simple extension of the humanistic basis for equal rights. To extend the domain of rights and give every citizen an equal share of the national income would give added recognition to the moral worth of every citizen, to the mutual respect of citizens for one another, and to the equivalent value of membership in a society for all."[3]

One must profoundly pray that humanism consists of something far more than such vaporous stuff. But even if one accepts Okun's notion of the humanistic base for equal rights at face value, it is difficult to perceive how he can overlook its obvious inconsistencies. Giving every citizen an equal share of the national income means either that every citizen contributes equally at the margin to the national income—an obvious impossibility unless income equalization in fact renders us all same, drab, and uniform—or that some must suffer expropriation so that bounties may be conferred on others. If we transfer income willingly to others, we may be perceived by virtue of the transfers as recognizing the moral worth of the transferees (although more realistically we are gratifying ourselves by making the transfers). If the transfers are unwilling and exacted from us by the state, in what way does this represent our respect for the transferees? Even more critically, how does this show respect by the transferees for the transferors? How are the rights of the transferors respected? How is the transferor's moral worth rather than his unworthiness reflected in the exaction?

Okun's humanism rests implicitly on a devout belief in the fantasy of a free lunch; the poor may be made more affluent without the affluent being made poorer by a redistribution policy. Egalitarians like to play out this fantasy by pretending that alleviating poverty figures in some social utility function. Since private charity clearly appears in private utility functions, surely we are all collectively prepared to pay something for collectivized charity. Even if we are prepared to make the heroic leap from private utility functions in which one of the arguments is voluntary transfers to the poor to some social utility function with nonvoluntary transfers to various groups, the analytics involved in determining optimum transfers and the tax schedule to effect them are horrendous (fanciful is too kind a characterization).

I do not gainsay Okun's or any egalitarian's preferences. Tastes in matters of aesthetics and morality vary; and they do not lend themselves to rigorous analytical determination. But they are subject to challenge and opposition, particularly if it is proposed that they be implemented in public policy. The enormous economic wrenching that equalizing public policies would entail should be based on something far more substantial than the preference of one of our intellectual stars. If some of your property rights are to be expropriated to subsidize some anonymous persons—not of your selection, not because you deem them in need of your assistance, not in amounts that you believe to be appropriate to your perception of their needs—surely the basis for this enforced transfer should be something more persuasive and compelling than that the resulting distribution of property rights pleases someone else.

This takes me to the conclusion that adverting to ethical and aesthetic considerations dictates no unique pattern of income distribution and specifies no particular degree of progression in tax liabilities nor indeed any given configuration of the tax system. This conclusion abstracts entirely from the far more difficult question whether any given tax configuration can indeed accomplish some desired distribution of economic status. Rather, it simply notes that it is a visceral, not a logical process, that leads one to egalitarianism; unless there is reason to rely on the viscera of egalitarians rather than those of libertarians, the shape of the income distribution and the means for attaining it must be otherwise resolved.

I do not make this assertion to demean the earnest efforts of scholars to formulate rigorous and workable systems of social justice. Egalitarians have not monopolized the field; libertarians have also struggled with real-world situations recalcitrant with respect to the inherent "logic" of individualism, utilitarianism, and freedom. On the whole, however, the preferences embodied in libertarianism appear to conform far more closely with the economic results of virtually any social system, including the highly collectivist which disparately reward their members; egalitarianism, however lovely it sounds, is virtually everywhere rejected in practice. But this is the utmost in casual empiricism; my positivist bent precludes dwelling further on it.

This argument—that economic equalization in fact is equalization of consumption, an objective that can scarcely lie at the heart of social justice and which must be repugnant in an affluent, diverse society—certainly does not exhaust the inventory of objections that economics addresses to egalitarianism. It suffices, however, for my present purpose, to present a new challenge to the egalitarian premises so deeply buried in the constructs of public policy.

Economics: Is Income Redistribution Feasible?

The ethical-aesthetic structure of egalitarian policy is not likely to come tumbling down because of this argument. Egalitarianism as a motive force in public policy will survive this feeble onslaught as successfully as it has resisted others. Leaving ethical-aesthetic considerations to some other resolution, we must direct our attention to the second set of issues, which consists of matters of economics. The central question is whether the use of fiscal powers *can* be directed successfully at achieving some desired distribution of income.

My answer to this question is no, unless it is possible to devise a fiscal system that achieves a more nearly equal distribution of the ownership of capital without simultaneously discouraging rich and poor alike from acquiring capital. To achieve its objective without making society as a whole poorer, a redistributive fisc must somehow contrive to subsidize the holding of capital (human and nonhuman) and its accumulation and use by the poor without penalizing the holding, accumulation, and use of capital by the rich. The tax system in this fiscal structure would have to focus not merely on the income level of the individual, but on its source as well; capital income received by the poor would have to be taxed at a lower rate than that applied to such income received by the rich, if indeed it were not actually subsidized. Moreover capital income would have to be taxed no more heavily than other income.[a]

This answer derives from the following propositions:

The amount of one's income is a function of the amount of capital one owns and/or uses in production.

Hence the shape of the income distribution depends on the distribution of capital ownership and its allocation among alternative uses.

Changes in the aggregate capital-labor ratio result in changes in total output and income and in the payments per unit of capital and labor service, but they do not change the shape of the distribution of income.

The redistributive fiscal policies that have been pursued have disregarded these analytical tenets. As a consequence they have had little effect on the shape of the income distribution; their principal consequence has been to reduce the rate of growth of production capability. These policies have left us collectively poorer than we need have been while the Lorenz curve remains deeply bowed.

The reason is that traditional income-redistributive fiscal policies heavily tax the rich while directing a substantial fraction of expenditures toward subsidizing consumption by the poor. But the rich are rich because they own and claim the returns to a disproportionate amount of capital, both human and nonhuman. Irrespective of whether the present tax system is deliberately focused on disproportionately taxing the returns to capital owned by the affluent, it has the effect of raising the cost of obtaining future income relative to the cost of

---

[a]While the present tax system, by virtue of the graduated income tax rate structure and personal exemption and the graduated estate and gift taxes, does indeed levy less heavily on poor recipients of capital income than on rich recipients, this differential is swamped by the fact that for rich and poor alike the total tax on capital income is far heavier than that on other income.

consumption, rather than of reducing the cost of saving by the poor relative to the cost of saving by the rich. In turn, the consequence has been less total saving rather than a more even distribution of saving; less total saving is the same as a smaller total stock of capital than would otherwise have been accumulated. Total output and income, therefore, have increased less rapidly than otherwise, but the distribution of income has not become less unequal because the distribution of the smaller stock of capital has not been materially affected.

Capital Endowment and Income

The analysis leading to these conclusions turns on one of the most basic and elementary propositions of economics, the law of variable proportions or of diminishing returns. This proposition holds that as the quantity of one production input is increased, while the quantities of the other inputs are held constant, total output increases but at a decreasing rate.[b] The marginal product of the variable input thus decreases with increases in its amount. By the same token, since the sum of the marginal products of each factor times the quantity of each must just exhaust the total output, the marginal products of the other production inputs increase as the amount of the variable input increases.

The law of variable proportions is universal; it holds irrespective of time, place, and social or economic organization. The authority for this observation is derived from Genesis in the Old Testament. The Lord, omniscient as well as omnipotent, rested on the seventh day, not because he was exhausted by his previous efforts, but to decide just how Adam and Eve would earn their daily bread after they were evicted from the Garden of Eden, the land of abundance. He first determined that since there must be no more free lunch, there must be diminishing returns for each of the production inputs that the ingenious minds of Adam and Eve, having eaten the fruit of the tree of knowledge, might devise. And while he had no intention of precluding his children from marvels, miracles he kept as his own. Hence there were to be no increasing returns to scale; the best that might be achieved would be increases in total output in the same proportion as that of total inputs, when every one of the latter were increased in the same relative amount. By the same token, neither could there be decreasing returns to scale.

With this stricture against the free lunch, additional production inputs would be no more freely available than additional output. Scarcity—greater wants than means to satisfy them—was to be the lot of men and women. And with limited production inputs, the cost of using an additional unit of any one input for any production purpose clearly becomes the amount of some other production that is lost as a result of having one less unit of that input.

[b]This is a highly truncated statement of the law of variable proportions, referring only to that part of the schedule relating total output to the amount of the factor input that is realistic and relevant for decision making.

From these simple laws emerged the rule that Adam and Eve might follow to earn their bread, in all its various forms, most efficiently. They should use each production input for any production purpose in such quantity that its marginal product just equals its marginal cost. Obviously, if Adam and Eve might divert one more unit of a production input from output A to output B and gain more of output B than they would lose of A, it would pay for them to do so (assuming that marginal quantities of A and B were equally prized).

Finally the Lord devised the specific relationships between the amount of the change in production input and the change in total output. An important property of this production function is that the percentage change in output is a constant fraction of the percentage change in the input, when the other inputs are held constant. This basic production relationship was identified and measured by Professors Douglas and Cobb, many years ago at the University of Chicago.[4]

The Lord did not specify the way in which his children might organize economic activity, allowing them to fumble their way to a system that would, if allowed to operate, permit efficiency in the use of production inputs while allowing individuals freedom of choice, including the freedom to be wise or foolish. In this system owners of production inputs would receive as claims on the total output amounts equal to the marginal (value) product times the number of units of the production input supplied. Where this system operates, the share of the total output received by the owners of each production input is the same as the percentage change in total output resulting from a given percentage change in each input. Since the latter is constant, so is the former.

To repeat, the law of variable proportions tells us that if we increase the amount of capital inputs relative to other production inputs in the production process, total output will increase, albeit at a diminishing rate. The production function tells us by how much output will increase; for example, if the elasticity of output with respect to capital inputs is, say, one-third, then a 1 percent increase in capital, other inputs held constant, will result in a 0.33 percent increase in output. The production function also tells us how much of the expanded output will go to each factor of production, provided that each factor receives, per unit, its marginal (value) product. Thus, of the 0.33 percent increase in total output resulting from a 1 percent increase in capital inputs, owners of the capital will receive one-third, or 0.111 percent, while owners of other production inputs will obtain an additional 0.222 percent.

This simple arithmetic illustrates two important principles. First, an increase in capital relative to other production inputs increases the aggregate income of the other inputs more than it increases that of the capital. Second, this increase in the ratio of capital to other inputs reduces the income per unit of capital while increasing the per unit returns of the other production inputs. In this example the amount of capital inputs increases by 1 percent, but the total payments for capital increase by only 0.111 percent; the return per unit of capital falls by 0.029 percent. By the same token the return per unit of the other factors increases by 0.222 percent. For ease of exposition, let us call all the other factors labor. Then an increase in the capital-labor ratio increases the return per unit of labor while decreasing that of capital; the respective labor and capital shares of the enlarged total income, however, remain the same.

In an efficient, free-market system, the capital-labor ratio depends on the respective costs per unit of capital and labor services. Earlier we defined the cost of a production input in terms of the alternative outputs that must be foregone when it is applied to a particular use. At the aggregate level the cost per unit of capital service is the amount of consumption that must be foregone to have the amount of capital that provides a unit of service; the cost per unit of labor service is the value of the amount of leisure that must be foregone. The efficiency-maximizing rule tells us that the quantities of the inputs used in production will be such that their marginal value products equal their marginal factor costs. An increase in the capital-labor ratio increases the marginal (value) product of labor, hence increases the quantity of labor services that will be demanded at any given price per unit of such service. But as the quantity of labor service demanded increases, the price per unit of labor is likely to rise also. In the general case the ultimate market result will be that more labor services will be employed at a higher wage per unit.[c]

Of course the capital-labor ratio is not likely to increase unless the cost of capital services decreases or the productivity of capital increases. Technical progress—implemented advances in the state of the industrial arts—tends to reduce the cost of capital services; that is, it tends to reduce the amount of current consumption that must be foregone to have the capital that provides a given marginal value product. In a tax-ridden economy the cost of capital services may also be altered by changes in the tax system. Whatever the source of the change, reducing the cost of capital services, other things equal, leads to an increase in the capital-labor ratio.

On this score there is little disagreement. To be sure there are widely ranging opinions as to the relative contribution of increases in the stock of capital to increases in total output and to increases in labor's productivity, employment, and real wage rates. The question is whether these effects are consequential with respect to the shape of the distribution of income.

The Capital-Labor Ratio and the Shape of the Income Distributions

The preceding discussion shows that changes in the capital-labor ratio will effect total output and income (hence the amount of total labor and capital income),

---

[c]The general case is taken to be that in which the supply of labor services is less than infinitely elastic but more than zero elastic with respect to the real wage rate.

will not affect the respective proportionate shares of total income, but will change the return per unit of input. The question is whether these changes in per unit returns affect the shape of the income distribution.

The answer to this question, regrettably, must rely on heuristic rather than empirical analysis.[5] The data on the distribution of income by income level are of poor quality, in part because of the way in which the information is collected, in part because of ambiguities about what is and what is not income, in part because of questions as to the appropriate entity—the individual, the family unit, the household—whose income is to be measured, in part because of failure to adjust for differences in age of the entity, and, insofar as real rather than nominal income is the pertinent measure, in part because of the failure to make the complex adjustments from current to constant dollar flows. In the light of all of these deficiencies it is difficult indeed to have even well-advised inpressions about the shape of the income distribution, let alone reasonably reliable measurements.[d]

I cannot attempt, in the compass of this paper, to develop new empirical evidence about the shape of the income distribution, nor is it worthwhile for the purposes of this discussion to dwell on the deficiencies of the existing measurements. In view of the data limitations I shall not attempt to demonstrate empirically the effect of changes in the capital-labor ratio over time on the distribution of income. Instead I shall rely on an heuristic analysis to show that a change in the capital-labor ratio does not change the income-level distribution of income and hence that redistributive policies of the sort long pursued in the United States and elsewhere are ineffectual. In doing so I shall assume away the complications of the age distribution of the population and of the technical details of the concept of income.[e]

For purposes of this analysis income is taken to mean the payments received by the owners of production inputs for the use of the inputs in some given period of time. In long-run competitive equilibrium (when no production input is fixed in amount or specialized to a given production use), income per unit of input equals the sum of the equilibrium interest rate or rate of return on the cost of reproducing the capital instrument affording the unit of input plus the depreciation of the capital instrument in that period. In this context capital instruments include human as well as nonhuman capital.[f] In the short run

---

[d]It is clear, however, that these conceptual limitations have little bearing on the egalitarian's convictions that income is to a significant degree unequally distributed and that public policy can and should be addressed to reducing this inequality. The vigorous egalitarian thrust of tax policy is not based on solid empirical analysis with respect to either the magnitude of the problem at which it is aimed or the effectiveness of a redistributive tax structure. Indeed, for the most part, it is not based on analysis at all.

[e]Only the distribution of permanent income, however, is considered in the analysis; surely not even the most unreconstructed egalitarian would impose on public policy the requirement to level out transitory receipts.

[f]Very little of the total flow of production services in a given period of time represents "pure" labor; by the same token very little of the total income is payment for pure labor

payments received for the use of production inputs may exceed or fall short of long-run competitive equilibrium incomes, since the quantity of most production inputs allocable to a given production purpose cannot be instantly changed. This is clear for nonhuman capital; it is no less true, although possibly obscure, in the case of human capital which is the source of the specialization of labor. Thus, insofar as there is any "fixedness" of capital, human or nonhuman, for any period of time, both labor and capital incomes at any point in time are likely to differ from their long-run competitive equilibrium amounts.[g] The appropriate framework for concern with the distribution of income is clearly the long run; if redistributive public policies focus on the short run, they will interfere with the market adjustment processes that tend to eliminate the excess or shortfall of quasi rents from the long-run equilibrium returns per unit of production input. Attaining the desired distribution of long-run equilibrium income, therefore, is likely to be impeded by attempts to cancel the effects of disturbances on short-run quasi rents.

Within this context, consider some hypothetical unequal distribution of income. For example, suppose that 90 percent of the population receives 50 percent or $500 of the economy's total income of $1000; that 80 percent of their income ($400) is labor income and 20 percent ($100) is capital income; that the remaining 10 percent of the population also receives 50 percent of the total income; and that 53 1/3 percent ($266.67) of this group's income is labor income while 46 2/3 percent ($233.33) is capital income. Suppose also that a total of 500 units of labor service is supplied at a wage rate of $1.33 along with 2000 units of capital service at a price per unit of $0.167. Finally suppose that the amounts of labor and capital services supplied by each group in the population are proportional to the labor and capital incomes they receive.[h]

Now suppose that a 50 percent tax is imposed on capital income only. Owners of the capital will respond to the tax-induced reduction in their income by reducing their stocks of capital.[i] Pretend that the reduction in stock of

---

inputs. As a rough approximation, virtually all the total income flow may be perceived as payment for the services of capital, human and nonhuman. The returns to human capital, of course, are received primarily in the form of wages, salaries, and other forms of compensation for personal services. To simplify exposition these returns will be referred to as labor income, while the returns for the services of nonhuman capital will be labeled capital income; similarly human capital and its services will be referred to as labor, while the term capital is reserved for nonhuman capital.

[g]Any such difference, if expected to persist, leads to efforts to adjust the quantities of production inputs that will be provided in the future, that is, it results in additions to or decreases in the stocks of the capital affording the flow of production services.

[h]While the specific values for the variables in this illustration are not represented as realistic, the relationships among them, except for the percentage distributions of total income, are derived from an econometric model of the United States economy developed by Norman B. Ture, Inc., for analysis of the effects of tax changes on employment and income.

[i]Since any given amount of capital will now provide its owners less future income than before the tax was levied, the cost per dollar of future income—the amount of current consumption that must be foregone to acquire or hold it—has been increased; people will want to hold less capital at the higher cost, everything else being equal.

capital is instantaneous and that the reduced stock provides 1454 units of capital service, compared with the 2000 units before the tax was imposed and the stock of capital shrank. As the discussion of the law of variable proportions shows, one consequence of this reduced input of capital service will be an increase in the pretax return per unit of capital service—to $0.20 in this case, leaving an after-tax return per unit of $0.10. Similarly the wage rate will fall from $1.33 before the tax was imposed to $1.22. Total labor service input declines from 500 units to 478 after the tax is levied. The economy's total income falls from $1000 to $872 (478 × $1.22 + 1454 × $0.20).

If we assume that the full amount of the capital income tax revenues, $145 (0.50 × $0.20 × 1454), is paid as negative income taxes with respect to the labor income of the 90 percent of the population, this group's after-tax labor income is $494. Their capital income after tax is $44, giving them total income of $538, an absolute increase of $38 from their total income before the tax was imposed.

The pretax income shares of the 90 percent and 10 percent of the population remain the same as before the tax was imposed, 50-50. The after-tax shares, however, are altered. The lower 90 percent now receives 61.7 percent of total income while the upper 10 percent receives only 38.3 percent. But this leveling of after-tax income reduces the absolute amount of pretax income for both groups, reflecting (1) the reduced input and productivity of labor services; and (2) the reduced flow of capital services and their return.

Five conclusions emerge from this analysis: (1) The traditional redistributive fisc is unlikely to affect the pretax distribution of income because its effects, if any, on the distribution of capital are secondary to its effects on the total amount of capital. (2) Redistributive tax and expenditure structures may reduce inequality in the distribution of after-tax income. (3) To achieve any significant reduction in after-tax income inequality, the tax revenues must be devoted primarily to financing payments to the "poor." (4) A major cost of the income redistribution is likely to be a reduction in the amount of labor services employed (an increase in unemployment). (5) Any reduction in after-tax income inequality is likely to be attained at the expense of everyone's being poorer in absolute terms; most people will be little, if at all, better off in an absolute sense as a result of the redistribution of disposable income.

This illustration is overdrawn to be sure. First, the existing tax system levies, and levies heavily, on labor income as well as on capital income. Taken by itself, this tends to dampen the redistributive effect of the fisc. Second, less than 100 percent of tax revenues are distributed to the poor. This, too, would make the fisc less redistributive than the illustration suggests. Nonetheless the thrust of the illustration conforms reasonably well with reality. Capital income is taxed far more heavily than other income, and the preponderant part of government outlays does go to the poor.[j] The fiscal structure is strongly redistributive. The

---

[j] As measured in the national income accounts, total government outlays (federal, state, and local) were $531 billion in 1975. Of this amount $169 billion were transfer payments to

questionable evidence about the distribution of pretax money income shows no trend toward greater equality; while after-tax money income is less unequally distributed than pretax income, it does not appear to have become materially more equal over time despite the increase in the size of the redistributive fisc relative to the total economy. And as government has grown, the unemployment rate has tended to rise. The conclusions about the inefficacy of redistributive fiscal policy and its cost in terms of total real output and income appear to fit the facts of the real world quite well.

The Antisaving Bias in the Existing Tax System

This discussion seeks to show that traditional redistributive fiscal policy generally fails to reduce income inequality but succeeds in making us all poorer than we need to be. These results surely are not intentional; they must be attributed to misapprehension rather than malice.

Whether its results are intentional or not, the existing fiscal system of the United States unduly penalizes activities that increase income. On the expenditures side of the fisc, an increasing proportion of the rapidly expanding government outlays are transfer payments to the poor; most of these transfer programs have the effect of increasing the cost of effort relative to nonparticipation in production activity. On the revenue side of the fiscal ledger, the tax system greatly increases the cost of saving, hence the cost of holding and accumulating human and nonhuman capital relative to consumption. The response to this tax-induced increase in the cost of saving results is less total income than would otherwise be produced but relatively little reduction in income inequality.

These perverse consequences, paradoxically, are the outcomes of a fiscal policy that seeks precisely the opposite objectives, namely greater output and less income inequality. The misfocused fiscal structure is the product of either ignorance or neglect of the effects of taxes and expenditures programs on the relative costs confronting private sector entities. It is to these price effects of taxes, particularly on the cost of saving relative to current consumption, that this discussion now turns.

Every tax changes some relative price(s) or cost(s); indeed, if a tax is to be effective in accomplishing its principal objective—to transfer claims to resources

---

individuals; most of this amount went to low-income individuals. Purchases of goods and services totaled $339 billion, of which $179 billion was compensation of government employees and $160 billion represented purchases from the private sector. Of the latter amount roughly $110-$115 billion may be assumed to have been payments by government contractors for labor services. Thus about $290 billion of government purchases represented compensation for labor services. Assuming that the income-level distribution of this labor income is about the same as for the economy as a whole, a substantial proportion (conservatively, two-thirds) of total government outlays went to relatively low-income individuals, those with income less than say $9000-$10,000, and at least a third went to the very poor.

from the private to the public sector—it must raise some price or cost to private entities. A neutral tax system would raise all prices and costs in the private sector in equal proportion and hence increase private sector costs of all resource use compared to the costs confronting the government. By the same token a tax unneutrality or bias means that for private sector entities the tax changes the cost or price of some thing(s) relative to that of others.

The existing tax system in the United States is heavily biased against private saving, hence private capital formation. In very large part this bias against saving is inherent in the income tax because the tax is levied both on the portion of current income that is saved and also on the future income that saving purchases when the future income is realized. But this bias is compounded by the additional imposition of the separate corporate income tax. It is aggravated by the tax on realized capital gains and by estate and gift taxes. State governments also levy individual and corporate income taxes and inheritance taxes, adding substantially to the extra cost of saving compared to consumption. Local government property taxes also add significantly to the disproportionate burden on saving.

Income Tax. The simplest way of showing how the present tax system increases the cost of saving relative to consumption is with an arithmetic example. To begin with, let us focus on the bias against saving inherent in the present type of income tax.

First, consider an economy without any taxes and a person deciding how to use $1000 of income. He can purchase $1000 of consumption goods. Alternatively he can buy additional income for the future, that is, he can save the $1000. For example, he can buy a $1000 bond paying, say, $120 a year until maturity. Suppose the bond is due in 13 years and pays 12 percent interest per year. The person's total annual return on the bond may be perceived as $155.67—$120 of interest received plus the $35.67 set aside by the bond issuer to accumulate, at 12 percent, to $1000 over 13 years in order to repay the principal at maturity.[k] If the person is just at the margin of indifference between the two alternatives, then for him the real cost of $1000 of current consumption is the foregone $155.67 per year of additional income. Similarly the cost of $155.67 per year of additional income is the $1000 of foregone consumption; the cost per dollar of future income is $6.42.

Now suppose an income tax of the same basic configuration as the federal income tax is imposed at a flat rate of 48 percent. After paying the tax, the person can now buy $520 of consumption goods or $520 of bonds which, at the same 12 percent yield, provide an additional $80.95 a year of income—the $62.40 of interest paid each year and the $18.55 accumulating in the debt issuer to repay the principal at the end of 13 years. But the $62.40 of interest the

---

[k]As seen by the bondholder, the future return on his $1000 bond purchase is the $120 of interest per year plus the $1000 principal repayment at the end of the year thirteen, but $1000 after 13 years is the same as an annuity of $35.67 per year for 13 years, accumulating at the rate of 12 percent.

bondholder receives also bears the tax, leaving the person $51.00 after tax—$32.45 in after-tax interest plus the $18.55. With this tax, then, the person must forego $520 of consumption for $51.00 of additional income per year; the cost per dollar of the future income is now $10.20, or 59 percent more than in the absence of the tax.

If this basic bias against saving is to be avoided, either the amount saved out of current income must be exempt from tax, while the full amount of the gross returns are included in taxable income, or the gross returns to the saving must be exempt, while fully taxing the amount saved.[1]

Capital Consumption Allowances. Now suppose that instead of buying a bond, the person buys a piece of real capital, say a machine tool. Suppose the machine tool is expected to remain in use for 13 years; ignoring any difference in risk, assume that the person discounts its future income stream at 12 percent also. If there were no income tax, the person would deem the investment to be just warranted if it were expected to generate a cash flow of $156 per year per $1000 of the machine tool's cost.[m] In other words, with this cash flow (and no

---

[1]The proof may be presented in a simple notational exercise. In the absence of a tax, the equilibrium amount invested would be such that

$$C = Y^* \tag{1-1}$$

where C is the amount paid to acquire the capital generating a future income stream and Y^* is the present value of the future income stream. Clearly C is the amount of consumption foregone to acquire the future income.

Under the present tax law, with no change in the real yield on the capital, to acquire the same pretax future income stream Y^* requires foregoing $C/(1-t)$ of current consumption. But Y^* is also taxed. Hence

$$C/(1-t) = Y^* (1-t) \tag{1-2}$$

where t is the marginal tax rate. If saving were deductible, we would have

$$(C - tC)/(1-t) = Y^* (1-t) \quad \text{or} \tag{1-3}$$

$$C = Y^* (1-t) \quad \text{or} \tag{1-3a}$$

$$C/(1-t) = Y^* \tag{1-3b}$$

In other words, to have net of tax Y^* of future income would require $C/(1-t)$ of current pretax income. To have C of current consumption would also require $C/(1-t)$ of current pretax income. Hence the deduction of current saving with full taxation of the gross return would raise the cost of future income in the same proportion as the cost of current consumption. Comparison of (1-3a) and (1-3b) shows that currently taxing saving while exempting the returns (1-3b) is precisely the same as exempting saving and fully taxing its gross returns (1-3a).

[m]This is obviously the same gross return per year as in the case of the bond. In this case, however, there is no repayment of principal at the end of 13 years since the real capital is presumably completely exhausted at that time. Disregarding the rate at which the capital is actually used up in the production process, provision for its replacement could be made by reserving $35.67 of its gross return each year in a replacement fund; at the end of 13 years accumulating at 12 percent, this will amount to $1000.

tax) the person would be just at the margin of indifference between $1000 of current consumption and saving $1000 to invest in additional tools. The cost per dollar of future income is $6.42.

With an income tax at 48 percent, however, the person has only $520 of the $1000 of pretax income to invest. If no allowance for the recovery of the $520 initial amount saved and invested in the machine tool were provided in the tax law, the individual's after-tax return on the $520 saving would fall to $42.09 per year; the cost per dollar of the future income available to him would increase to $12.35, almost 1.9 times the cost in the absence of the tax. In fact, the present income tax laws do permit annual deductions for depreciation that moderate the effect of the tax on the cost of future income. For example, with straight-line depreciation based on a 13-year life of the machine tool, the after-tax gross return on this investment is $61.30. The cost, the amount of foregone current consumption, per dollar of disposable future income is $8.48; while this is 32 percent more than in the absence of the tax, it is obviously substantially less than if no capital recovery allowance were permitted.

The present depreciation provisions permit more rapid capital recovery than is afforded by straight-line allowances based on useful life. Both the formulas for computing the annual deduction and the asset depreciation range system (ADR), which allows depreciation to be computed on the basis of lives as much as 20 percent shorter than the standard, result in the deduction of substantially larger fractions of the investment in the early years of use of the capital than is permitted under straight-line depreciation with standard useful lives. And the investment tax credit supplements these so-called accelerated capital recovery provisions by allowing the investor a credit against tax, rather than a deduction against gross income, equal to given percentages of the investment. These provisions, taken together, further abate the adverse effect of the tax on the cost of future income derived from durable nonhuman capital. Using our example, suppose we have an investment credit at 7 percent, the double-declining balance depreciation method, and shortened equipment lives for tax purposes (such as the asset depreciation range system allowing the person to base depreciation deductions on a 10-year rather than 13-year life). If the person took full advantage of these provisions, the extra cost of saving compared with consumption would be 11.2 percent.

Provisions such as the ADR, the accelerated depreciation method, and the investment credit almost invariably appear high on the traditional tax reform list of loopholes, tax shelters, or tax expenditures. These tax provisions do, indeed, moderate the adverse impact of the income tax on the relative cost of saving invested in capital subject to depreciation; they also reduce, initially, the cost of future income derived from such capital relative to that obtained from other types of capital. They do not, however, provide for complete tax neutrality between such saving and consumption. To do so, the present value of the aggregate capital recovery allowances would have to be just equal to the present value of the outlays for acquiring the capital assets. This result clearly would be obtained if such outlays were immediately expensed for tax purposes.[n]

In this context evaluation of what used to be the number one loophole on the conventional tax reform list—percentage depletion for oil and gas properties—produces interesting results. The standard tax reform ploy was to cite the fact that since these allowances were related to the income generated by, rather than the cost of, depletable properties, the cumulative amount of such allowances might well exceed, possibly substantially, the total investment. Moreover, it was pointed out, much of the investment in such properties could be immediately written off by virtue of the deductions for intangible drilling costs. But examination of actual experience in the petroleum industry revealed that in the general case the present value of all these deductions, intangibles and depletion, was just about equal to the cost of the properties.[6] In other words these deductions resulted in the income tax's increasing the cost of future income obtained from oil- and gas-producing properties in just about the identical proportion that it increased the cost of current consumption; the tax treatment of such investments represented one of only a few cases of income tax neutrality with respect to saving and consumption of current income. (With the general principle that tax neutrality between current consumption and saving of current income requires either the deductibility of saving or the exemption of the return, the reader might want to evaluate some of the other "glaring loopholes," such as tax exemption of the interest on municipal debt instruments and the noninclusion of imputed rental income on owner-occupied housing).

At the opposite extreme is a considerable part of the saving invested in human capital. The individual who reserves some current income or who foregoes current income and consumption to pay for education and training to enhance his productivity is investing in human capital of varying durability. In the general case he is allowed no deduction either for the saving or for any significant part of the returns. The tax-induced increase in the cost of the future income produced by this capital is much the same as in the case of investment in the machine tool where no capital recovery allowance is afforded.

Corporate Income Tax. Separately taxing the income generated in corporate businesses significantly increases the tax bias against saving insofar as that

---

[n]Referring to the notational proof in footnote 1, we may rewrite (1-2) to take account of capital recovery allowances as

$$C/(1-t) = Y^* (1-t) + tD^*, \qquad (1\text{-}2a)$$

where D^* is the present value of capital recovery allowances. If saving invested in capital now subject to such allowances were deductible in the year in which the capital was purchased, there would of course be no capital consumption allowances and we would have,

$$(C - tC)/(1-t) = Y^* (1-t) \qquad (1\text{-}3)$$

The remainder of the proof is as shown in the proof in footnote 1.

income is taxed at a higher rate than that paid by individuals and insofar as the income is distributed and taxed to individual stockholders. To pursue our illustration, suppose a person buys stock in a corporation that uses the funds to buy the machine tool. Suppose the corporation, too, is subject to a 48 percent income tax. If the individual were content to leave his share of the corporate earnings in the corporation forever, and if no individual income tax liability accrued with respect to such retained earnings, the tax results would be the same as those already described. If he insisted on full distribution of his share of the after-tax earnings, however, he would pay an additional tax on the dividends he received. On his $520 saving he would obtain $31.87 after corporate and individual income taxes. The cost per dollar of future income, on these assumptions, rises to $16.32, compared with $6.42 in the absence of tax. The corporate income tax combined with the individual tax increases the cost of saving relative to consumption by 154 percent, when corporate after-tax earnings are fully distributed.º If the investment credit, the ADR, and double-declining balance depreciation are used by the corporation for tax purposes, the increase in the relative cost of saving is still 113 percent.

The corporate income tax not only greatly augments the tax bias against saving, it also distorts the financial capital structure of corporations. This distortion results from the deductibility by the corporation of interest payments but not dividends for purposes of determining taxable corporate income. The magnitude of this distortion depends on the ratio of debt to equity financing and the actual marginal tax rates of the corporation and the individual investor. If we use the assumptions of the previous example and in addition assume the corporation pays 12 percent a year on its indebtedness, the cost to the individual per dollar of future income is $16.32 in the all-equity financing case, $11.79 where financing is half debt and half equity, and $10.20 in the all-debt case; the tax-induced increase in the cost of future income is, respectively, 154 percent, 84 percent, and 59 percent. Table 1-1 shows how these results are obtained.P

ºTo be sure, full distribution of after-tax cash flow is a rarity. Suppose the corporation distributes 25 percent of gross earnings less tax. Then the individual investor nets $53.94, including in this amount the earnings retained by the corporation. The cost per dollar of future income is $9.64; this is 50 percent more than the cost in the no-tax world.

PThis example ignores the effect of differences in debt-equity ratios on the yield that investors will require on both debt and equity investments. Since service of debt exerts a prior claim on a company's financial resources, holders of the equity interest in the company are at greater risk the greater the ratio of the company's debt to its equity. In real life therefore a corporate management seeking to maximize the value of the equity interest in the company confronts a financing trade-off. On the one hand, it may reduce the cost of providing future income by using debt issues to finance capital additions; on the other hand, it may encounter an increase in the amount of future income per dollar of equity investment that will be required to maintain the market value of the equity as the perceived risk of an increasing debt-equity ratio raises the applicable market discount rate. The differences in the cost per dollar of future income with respect to the alternative methods of finance are accordingly likely to be less than the example suggests.

Table 1-1
Tax Impact in Corporate Financial Structure

	Financing		
	All Equity	Half Equity, Half Debt	All Debt
(1) Investor's after-tax income-foregone consumption	$520.00	$520.00	$520.00
Corporation			
(2) Pretax gross earnings (15.57 percent per year for 13 years)	80.95	80.95	80.95
(3) Depreciation deduction (straight line)	40.00	40.00	40.00
(4) Interest deduction	–	31.20	62.40
(5) Taxable income (2) – (3) – (4)	40.95	9.75	–
(6) Tax (48 percent)	19.66	4.68	–
(7) Reserved for debt repayment[a]	–	9.28	18.55
(8) After-tax gross earnings less interest and reserve for debt repayment (2) – (6) – (4) – (7)	61.29	35.79	–
(9) Dividend	61.29	35.79	–
Individual			
(10) Dividend income (9)	61.29	35.79	–
(11) Interest income plus income reserved for debt repayment (4) + (7)	–	40.48	80.95
(12) Taxable income (10) + (11) – (7)	61.29	66.99	62.40
(13) Tax (48 percent)	29.42	32.16	29.95
(14) After-tax income (10) + (11) – (13)	31.87	44.11	51.00
(15) Cost per dollar of income (1) – (14)	16.32	11.79	10.20

[a]Annual amount which, accumulating at 12 percent, equals principal amount of debt to be repaid at end of year 13.

Capital Gains. It is ironic that the income tax provisions widely identified as one of the principal tax breaks for the affluent—the capital gains provisions—are in fact major sources of the tax bias against saving. Some capital gains are illusory, that is, the results of inflation; in addition, taxing both the capitalized amounts of future income and the future income stream itself clearly must increase the cost of future income compared with current consumption. The view that capital gains are income, only insignificantly different from other forms of income, hence properly taxable as ordinary income, is a product of the treatment of changes in stocks, reflecting anticipated or actual changes in flows, the same as the changes in flows themselves.

Apart from changes in the value of assets attributable to changes in the price level, capital gains are the market measure of the capitalized value of expected increases in income generated by the assets. Such increases in income, in the ordinary course of events, will be taxed as they materialize over time. To tax their capitalized value is obviously to tax the same future income twice.

These anticipated increases in the income provided by the assets may result from valuation changes, from increases in the market demand for the goods and services the assets produce. They may, however, merely reflect current saving and the future income expected from it.

This may be illustrated by reference to the preceding example. Suppose that the corporation retains its after-tax earnings, (saves them) and invests them in equally productive machine tools. For each year's retained earnings of $61.29 (in the all-equity case), annual pretax gross earnings are $9.54. If the investor sells his stock in the corporation at the end of the first year, he obtains $562.78 (the present value of $80.95 for the remaining 12 years of life of the machine tools purchased with his initial $520 investment plus $61.29, the present value of the additional earnings of the corporation). Ostensibly he has a capital gain of $42.78; under the present provision for including only 50 percent of the gain as taxable income, he pays a tax of $10.27; this is an additional, one-time $10.27 tax withdrawal from the returns to the $520 of saving. Its effect is the same as adding $1.43 of tax per year over the life of the asset. It increases the cost per dollar of the future income from $6.42 in the taxless world to $8.69, an increase of 35 percent. The capital gains "loophole"—taxing such gains at marginal rates equal to half the rates applied to "ordinary income"—turns out to be negative.

State Income Taxes. Individual income taxes are levied by 41 states and corporate income taxes are imposed by 45 states. These state income taxes generally conform closely with the federal income taxes. In some cases the federal income tax is deductible for purposes of determining the tax base for the state's income tax; state income taxes are deductible for purposes of determining taxable income for federal income tax purposes. Notwithstanding the latter deductibility, the imposition of state income taxes further increases the bias against saving since these taxes, like the federal, tax both the amount saved and the returns. Although the increase in the cost of future income relative to the cost of current consumption resulting from these state income taxes may be relatively modest, because of the substantially lower rates at which they are levied, it is far from inconsequential. For example, if the effective rates of state corporate and individual income taxes were each 4 percent, the cost of a dollar of future income in the all-equity financing, full-distribution case would be $18.16, about 183 percent more than in the taxless world and more than 11 percent more than if only the federal income taxes were applied.

State and Local Government Property Taxes. State and local government property taxes significantly add to the disproportionate cost of saving compared with consumption. While these taxes are ostensibly levied on the market value of property, they are in effect imposed on the income stream generated by the property. If the effective rate of the tax, that is, the rate when measured against the "true" market value of the property, were 4 percent and the annual gross yield on the property were 15.56 percent, as in the example we have been using, then the tax is the equivalent of a 25.7 percent levy on the gross yield of the property. To pursue this example (again, using the all-equity financing, full-distribution case), and allowing for its deductibility in the computation of taxable income for both federal and state income taxes, we see that the property tax increases the cost per dollar of future income to $21.64. This is about 237 percent more than in the no-tax world and more than 19 percent more than in the absence of the property tax.

Estate, Gift, and Inheritance Taxes. Federal and state taxes on property transfers by gift or at death are akin to capital gains taxes with respect to their effects on the cost of future income compared with present consumption. The base of such taxes is the market value of the transferred property, which in turn equals the present value of the future income the property is expected to produce. That future income is taxed as it materializes over time. Taxing its capitalized amount on the occasion of the property transfer is an additional tax levy on the same income stream.

Moreover the property may also be perceived as the accumulated amount of past income reserved from consumption. Again that past income was taxed as it was received. Taxes on the value of the property on the occasion of its transfer are another layer of tax on the same income stream. The extra burden of these transfer taxes on saving is mitigated by the various tax provisions that reduce the amount of the taxable property. It is also moderated by the remoteness in time of the tax liability for many individuals; the present value of the tax liability, as it enters saving-consumption choices, is likely to be quite low except for the elderly or those contemplating inter-vivos transfers in the relatively near future. Notwithstanding, these taxes must be seen as incremental burdens on the returns to saving, hence as increasing the cost of saving relative to current consumption.

The Aggregate Antisaving Tax Bias. This arithmetic example does not purport to be realistic; it is presented only in the interests of explaining how various components of the tax system distort the relative costs of current consumption and future income. On the one hand, few taxpayers have incomes so large that they face marginal income tax rates as high as 48 percent; moreover a large inventory of shelter arrangements reduce the marginal tax rates for many of the relatively few persons whose incomes would place them in the upper tax brackets. On the other hand, the example relies on a very short time span for receipt of future income, only 13 years. If one assumes more realistically that the saving-consumption choice pertains to, say, 30 years of future income, the tax distortion is substantially greater. To illustrate, we return to the first part of the example in which the individual purchases a bond. If the term of the bond were 30 years with an interest rate of 12 percent, the cost per dollar of the future income in a no-tax world would be $8.06. With a 48 percent tax, this cost increases to $15.03, an increase of 86.6 percent, in contrast with the 59 percent increase for the 13-year bond.

With all federal, state, and local taxes taken into account, the tax-induced increase in the cost of future income relative to consumption was about 104 percent in 1976, before enactment of the 1976 tax reform legislation that augmented the antisaving bias of the income taxes.[q] In effect the overall tax system of the United States imposes an enormous excise on saving compared with consumption of current income. This excise is moderated with respect to some forms of saving, but similarly it is substantially larger than the indicated rate for others. These differentials must surely affect the allocation of saving among capital uses in the same way that differential excises on consumer goods must distort the composition of consumption.

---

[q]The tax-induced percentage increase in the cost of future income X is defined as

$$X = 1/(1 - t_K) - 1 \tag{1-4}$$

where y is the marginal gross yield of capital, and t_K is the marginal rate of tax on the gross returns to saving. Expression (1-4) is derived as follows. In the absence of taxes a dollar of pretax income buys either a dollar of consumption goods or $\$y$ per year of future income; the cost of a dollar of future income is then $\$1/y$, which is the amount of current consumption that must be foregone. With an income tax of the present configuration, a dollar of pretax income buys $\$(1-t_T)$ of current consumption where t_T is the overall marginal rate of tax on all income; the cost of a dollar of consumption is then $\$1/(1-t_T)$. With the tax a dollar of pretax income buys only $\$y(1-t_K)(1-t_T)$ of future income; the cost of a dollar of future income is then $\$1/y(1-t_K)(1-t_T)$. The relative cost of the future income is

$$\frac{\$1}{y}\left(\frac{1}{(1-t_K)(1-t_T)}\right) \div \frac{1}{(1-t_T)}.$$

The increase in the relative cost of future income is

$$\frac{1}{y}\left(\frac{1}{(1-t_T)(1-t_K)}\right) \div \frac{1}{(1-t_T)} - \frac{1}{y}$$

and the percentage increase is

$$X = \frac{[1/y]\,[1/(1-t_T)(1-t_K)] \div 1/(1-t_T) - 1/y}{1/y} \tag{1-5}$$

which simplifies to (1-4).

Based on national income data, the estimated marginal rate of tax—federal, state, and local—on the gross returns to saving was about 51.0 percent. The tax-induced extra cost of future income relative to current consumption, expressed in percentage terms, is about 104 percent.

In Defense of Loopholes

It is to these allocational distortions that the conventional tax reformer addresses his reform proposals. The economic analysis provided in support of such proposals appears to confirm the equity arguments, namely that these differentials in the effective rates of tax on capital income, "loopholes," result in unequal tax liabilities for individuals (or companies) with equal incomes and hence violate the horizontal equity standard; that they are available primarily to rich individuals (and big companies) and hence violate vertical equity criteria; that they also induce an allocation of capital different from that which would result in tax-free markets and hence violate the neutrality criterion. Closing loopholes, therefore, makes the tax system fairer and reduces its effect on the allocation of capital and other production inputs, and hence on the composition of output. Moreover closing loopholes would raise substantial amounts of additional revenue that would permit reducing nominal tax rates without net loss of revenue. Not only would the tax system be fairer and more nearly neutral, it would create fewer incentives to find new loopholes.

This all sounds highly persuasive. Surely, however, it is a peculiar way of ordering priorities. Suppose the tax structure included a system of excises on consumption goods with an overall rate of about 100 percent. Would we identify as a loophole some feature of the excise system that allowed a particular group of consumer goods, say meats, to bear a tax of only 50 percent? Would we argue that good tax policy calls for increasing the rate on these goods to 100 percent so that we might lower the overall rate to, say, 75 percent? If we find an overall rate of 100 percent too high, why not attack that problem directly? If we were to do so, we would find that the lower effective rates on some consumer goods were less consequential as loopholes. The inequity of these lower rates would be moderated and the resource misallocations they induce would be similarly mitigated. Why should we follow the conventional tax reformer's hilly route to the same ends?

The answer lies, I suspect, in the conviction that the ultimate objective is not to reduce overall nominal rates of tax on capital income but to use the (allegedly) substantial additional revenues from closing loopholes either to finance additional government spending and/or to reduce effective tax rates (at least relative to what they otherwise would be) for low-income individuals. Recent history appears strongly to support this surmise. The tax legislation of 1969, 1971, 1975, and 1976 variously reformed and reduced taxes. Most, indeed virtually all, of the reforms involved increasing the effective rates of tax on one or another capital income stream; most of the tax reductions were focused on low-income individuals.

I do not mean to argue that we should be complacent about the capital-allocation distortions exerted by the present tax structure. They do indeed involve efficiency (welfare) losses. But such losses must be deemed to be of relatively small consequence compared with those from taxing capital formation uses of production capability in the aggregate 100 percent more heavily than consumption uses.

Indeed, until evidence of a new, more constructive perspective on these issues by policymakers emerges, there is much to be said for efforts to enlarge existing loopholes and to create new ones to enable capital income to avoid the full fury of the presently punitive tax system. There is, after all, a highly efficient capital market which operates to equalize risk-adjusted net returns to all capital uses. Unless it can be demonstrated that the capital markets consist of airtight, specialized compartments precluding the flow of saving among alternative uses, cutting the tax on the returns to any given use of capital must be diffused to all capital uses. While the owners of the capital initially relieved by the tax cut may *appear* to be the sole beneficiaries of the relief, they must in short order share these benefits with all who provide capital services.

Reducing the effective rate of tax applicable to the returns to any one use of capital not only increases the net returns on that use, thereby attracting additional capital to it, but also reduces the overall extra tax burden on all returns to saving. This must induce a shift in the allocation of current income from consumption to saving unless people are utterly indifferent to the relative cost of saving in making their saving-consumption choices.[r] With more of the current income saved, hence more of current production capacity used to add to the stock of capital, future income will be larger than otherwise. The preponderant part of the additional income goes not to the owners of the additional capital but to those providing the other production inputs with which the capital is used.

The analysis is symmetrical; just as opening loopholes reduces the overall cost of saving relative to consumption, closing them increases the relative cost of saving. Unless loophole-closing legislation is inextricably tied to nominal tax rate cuts on capital income sufficient *at least to cancel the increase in the relative cost of saving from loophole closing,* even the balanced tax-reform, tax-cut package must have the perverse result of eroding capital formation. As we have seen, a slower rate of growth in the stock of capital means retardation in the growth of total income, primarily noncapital income. Loophole closing is a costly business.

The phrase in the preceding paragraph is italicized to emphasize that equality of revenue effects as conventionally measured is *not* the appropriate

---

[r]The presumption that saving is zero elastic with respect to its relative cost requires that the demand for consumption be more than zero elastic with respect to its cost. Suppose for example that the cost of saving increases 10 percent relative to the cost of consumption, that is, that the cost of consumption falls by 9.1 percent. By hypothesis the amount of future income purchased is unchanged. Then the amount saved must *increase* by 10 percent and the amount consumed must *fall* in an equal dollar amount. This produces the absurd result that as the cost of consumption decreases, so too does the quantity of consumption. Surely it makes far more sense to assume that the quantity of future income demanded decreases (increases) as the price per dollar of future income rises (falls).

index of the effects of tax legislation. The standard measure of revenue effect assumes no behavioral response to tax changes; it ignores entirely the shifting and ultimate incidence of tax changes. At most, such revenue estimates attempt to take into account changes in total income through time, where such changes are not affected by the changes in relative costs and prices and the responses resulting from the tax changes. The interpretation and utility of such revenue estimates elude me. The only thing that may be confidently said about such estimates is that they are certainly wrong. They could be reasonable approximations only on the impossible assumption that all price elasticities are zero. When we are dealing with proposed legislation packages ostensibly designed to have some specified effect on the allocation of resources, for example the allocation between capital formation and consumption goods production, the conventional revenue estimates offer us no useful information to this effect.

What we need instead is a translation of proposed tax changes into their initial effects on relative costs and prices and an analysis of the responses by private sector entities. These responses, assuming they can be estimated with reasonable accuracy, tell us the allocative effects of the tax changes. Similarly they tell us the changes in the magnitude and composition of the tax bases, hence the changes in the amount and distribution of tax liabilities. We must look to these net revenue changes if we are to assess the effects of tax legislation on the distribution of tax burdens and after-tax income.

This is a large order to fill. To do so requires the development of an econometric model that, at a minimum, specifies and estimates the technical production relationships in the private and public sectors, the conditions of supply of production inputs to their alternative uses, the conditions of demand for various private sector outputs, the processes of adjustment to changes in relative prices, including those induced by tax changes, and the income-level distributions of the suppliers of production inputs. Models that do not satisfy these minimum requirements either fail to inform or, more likely, misinform.

In this context the standard list of loopholes, or tax expenditures, the current euphemism, should be viewed with a jaundiced eye. The justification for inclusion of any number of tax provisions on the usual tax expenditure list must raise eyebrows in any case; what argues, for example, that the corporate surtax exemption is a tax expenditure? The fundamental question, of course, is what concept of income or other relevant tax base appropriate for tax purposes serves to identify tax provisions as tax expenditures? Boris Bittker has addressed this question in a devastating article, "Accounting for Federal 'Tax Subsidies' in the National Budget" (*National Tax Journal* 22, 1969, pp. 244-261). My discussion takes a somewhat different approach from Bittker's but similarly challenges many of the conventional views about loopholes. Thus, when interpreted (as I believe they should be) in terms of their effects on relative costs, provisions such as the 50 percent deduction for long-term capital gains, the ADR, and various accelerated depreciation formulas turn out to be at best mitigations of differentially heavy excises on the purchase of future income rather than tax loopholes. But even if one rejects Bittker's or my challenges about the identification of tax expenditure items, one must still be highly skeptical about the conventionally measured amounts of these items.

It is regrettable that tax reform efforts are based on misapprehensions regarding the character of various tax provisions and of the effects of changing them. It is even sadder that these efforts rely on and are guided by unrealistic quantitative estimates of their effects. To a substantial extent the targets of tax reform are provisions deemed to benefit primarily the affluent. The magnitudes of such benefits provided policymakers are the conventional initial impact revenue estimates; so, too, are the increases in the tax liabilities of the rich, hence the decreases in their disposable incomes. With such grossly misleading quantitative guides, it is not surprising that redistributive tax policy accomplishes so little.

An Effectively Redistributive Tax System

The egalitarian thrust of the tax system is aimed at capital. The principal consequence of this focus is to reduce the aggregate stock of capital below the levels that might otherwise be achieved. Society as a result is poorer, but its members are not materially more nearly equal in economic status. The final question to which this discussion is addressed is whether it is feasible to construct a tax system that contributes to greater economic equality without unduly reducing society's total wealth and income.

I think it may be. This answer is deliberately tentative since the tax system I propose, while not new in concept, would require a draconian shift from the conventional ideology of taxation. In the abstract the proposed tax system is feasible; it would result in greater levels of production capacity and income than are forthcoming under the present tax system, and it would increase the wealth and incomes of the poor relative to those of the rich. Doubts about its feasibility, therefore, pertain principally to its political acceptability to egalitarians. I would be less diffident on this score if there were some evidence that egalitarians were beginning to see the forest, not merely the trees.

The tax system required for the stated purposes must be more nearly neutral than the present one in its effects on the relative costs of saving and consumption. It must also contrive to reduce the relative cost of saving for the poor to a greater degree than for the rich, so that a larger proportion than at present of an enlarged stock of capital is owned by the poor. Let us consider these requirements separately.

Tax Changes to Reduce the Bias against Saving

Complete elimination of the present tax bias against saving would require drastic changes not only in the federal tax system but in the tax structures of state and local governments as well. The property tax on which local governments so heavily depend is strongly biased against saving. Unless the revenue independence of local governments is to be sacrificed or unless some alternative revenue sources less punitive of saving yet adequate for local government revenue demands are devised, a significant antisaving element would remain in the total tax structure even if such elements in the federal and state tax systems could be entirely eliminated. This part of the discussion, accordingly, is concerned principally with alterations of the federal tax system.

At the federal government level the focus of tax changes would be primarily on individual and corporate income taxes; estate and gift tax revisions would also be highly desirable, although of secondary consequence. In addition, major and basic revisions in the present social security system are required to moderate, if not entirely to eliminate, the very large, adverse impact of the system on private saving and on the effort-leisure choice.[7]

Income Tax Revisions. The primary revision called for in the interests of greater income tax neutrality with respect to saving and consumption is to exclude from the base of the tax either the amount of net saving in the taxable year or the returns on saving realized in that year. As a practical matter, since the returns on human capital are included in wages, salaries, and other compensation for personal effort, it is difficult to disentangle these returns from the payment for labor services. The deduction of saving from gross income with inclusion of all the returns on saving in the tax base is therefore more feasible than taxing current saving and exempting the returns.

Fully implementing this approach would require substantial changes in the computation of taxable income. Apart from such obvious changes as elimination of the present tax-exemption for interest received on state and local government debt instruments, this approach would require individual taxpayers to maintain complete net worth accounts in order to verify their saving each year. Such accounts would be needed to measure the difference between the taxpayer's gross income and his consumption outlays over the taxable period. This difference, his net saving, is approximated by his purchase of assets, irrespective of their form, less any increase in his indebtedness, irrespective of its form, that is, by the change in his net worth.

The greatest difficulty in this approach pertains to saving allocated to the accumulation of human capital. In terms of the simple tax accounting mechanics, the human capital component of an individual's true net worth is difficult if not impossible to evaluate; changes attributable to the individual's reserving some of his current income from consumption uses would be in many cases impossible to measure. Amounts explicitly paid for education and training would pose no problem; they would be treated precisely the same as amounts paid for stocks, bonds, real property, and so forth. But outlays to acquire other kinds of human capital are often difficult to distinguish from consumption. For purposes of the proposed tax, arbitrary distinctions among such outlays would have to be made. Notwithstanding the resulting incompleteness of the accounting for saving invested in human capital, the proposed treatment would represent a major advance over the present law which for the most part provides no deduction for such saving nor for recovery of the capital against the income it generates.[8]

Another problem would arise in the treatment of outlays for consumer durables (including those for owner-occupied housing). These outlays are, for the most part, conventionally treated as current consumption, although it is obvious that they are better described as household investment, since the services provided by consumer durables are generated and consumed over extended time periods. If such outlays were treated the same as other saving, however, it would be necessary to impute annual returns to them. For example, the purchase of an automobile for cash—that is, without any increase in the individual's indebtedness in that year—would be deductible from gross income, but an annual gross return on this investment would have to be included in gross income each year so long as the individual retained the car. If the individual borrowed some part of the cost of the automobile, the amount of the debt would be added to his gross income in the year it was incurred, while the annual service of the debt would be deductible until the obligation was retired.

The same sort of tax treatment would be required with respect to owner-occupied housing. In the year in which the purchase was made the individual would deduct from gross income the difference between the purchase price and his mortgage indebtedness—his equity, in other words—but would have to include in gross income each year the imputed gross rental value of the house while deducting the annual service of the mortgage.

Imputing income to large aggregates of consumer durables, including housing, is conceptually feasible, but as a practical matter the imputation of an income flow to particular assets in the hands of particular taxpayers would be difficult indeed. In lieu of any such individualization of imputations of income to consumer durables, it would be necessary to rely on ratios of gross income to investment in such assets determined on a highly aggregative basis.

The desirability of this treatment would be found only in its conformity with that of other forms of saving. It would represent only a slight refinement, in practical terms, over the present tax treatment. Under present law outlays for consumer durables are not deductible, but neither are the imputed returns on this savings subject to tax (except that any gain realized when the asset is sold or exchanged is taxable). Taken in the large, therefore, present tax treatment closely conforms with the neutrality standard. Since the proposed tax would not allow deductions for outlays, none of the returns, including gains on the disposition, would be subject to tax.

---

[8]The real cost of acquiring human capital often includes the earnings foregone by virtue of an individual's undertaking education or training as well as the explicit outlays. For any one person the valuation of such foregone income is extremely difficult if not impossible. These foregone earnings, however, would not be subject to tax; failure to include them as elements of saving, accordingly, would not involve the double taxation that results under present law.

Other forms of saving on which explicit rather than imputed returns are realized would be far easier to handle. Purchases of corporate equities, bonds and other debt instruments (irrespective of the issuer), deposits in savings accounts, increases in the cash value of insurance policies, purchases of business assets in the case of unincorporated businesses, increases in cash balances, and so on would present relatively few difficulties for the vast majority of taxpayers.[t]

Since the net saving used to acquire assets would be deductible, there would be no capital recovery allowances on depreciable or depletable property. The gross returns on such property, less the costs of servicing any debt incurred to acquire them, would be fully included in taxable income.

Taxable income would also include the full amount, not merely the gain, of any proceeds from the sale, exchange, or other disposition of any assets, including the receipt of the repayment of the principal of any loans. The use of such proceeds to acquire other assets, of course, would give rise to a deduction from taxable income; full reinvestment of the proceeds would afford the taxpayer "roll-over" treatment, without the necessity or occasion for basis adjustment.

This approach to tax neutrality would require the complete elimination of any separate tax on income generated in corporations. Corporate earnings would have to be attributed to individual shareholders and taxed to them only under the individual income tax.[u]

A number of adjustments would have to be made in the measurement of the corporate-generated earnings to be attributed to the individual shareholders. Since corporate retentions are by definition saving and since saving would be excluded from the tax base, the earnings retained by the corporation would have to be deducted from the conventional measure of earnings per share. With the deductibility of saving in the year in which it is undertaken, there would be no capital recovery allowances; the conventional measure of corporate earnings, therefore, would have to be increased by the amounts now allowed as deductions for depreciation, depletion, and other capital consumption allowances. Similarly all proceeds, not merely the gains included in them, from the disposition by the corporation of any of its assets would be added to the conventionally measured earnings assigned to shareholders.

Capital gains and losses would simply disappear as elements of taxable income. Since the saving used to purchase assets would be deductible, all returns on such saving, including all the proceeds from sale or other disposition, would have to be included in income as they materialized.

---

[t]Valuation problems would be encountered when the purchases were other than arm's-length. Including increases in cash balances as components of net saving is conceptually essential but poses the practical problem of inputing gross income to cash holdings, similar to the imputation problems discussed in connection with consumer durables.

[u]It would probably be desirable to have the corporation act as a withholding agent for shareholders. This would be particularly useful for purposes of collecting the tax due on income attributable to nonresident alien shareholders.

The proposed tax would entirely eliminate the existing differential tax treatment of debt and equity financing of capital outlays. The deductibility of interest payments and the nondeductibility of dividends would reduce the cost of debt compared with equity financing. Under the proposed tax individuals purchasing either debt or equity issues would deduct such purchases; the returns they receive would be fully included in their taxable incomes. The tax, accordingly, would make no distinction between debt and equity financing regarding either the initial issue or the subsequent service.

Estates and Gifts. The present taxes on the transfer of property by gift or at death represent an additional layer of tax on saving. A tax system seeking to provide neutral tax treatment of saving and consumption should not impose any such tax. On the other hand, the proposed income tax requires the taxpayer to maintain a full account of his net worth; increases in his net worth equal his saving in the period in which they occur and are deductible, while decreases represent disposition of assets, the full amount of which is included in his taxable income. Under these rules the transferor of property by gift or bequest would have an increase in his taxable income in an amount equal to the gift or bequest; the property recipient, on the other hand, would deduct the amount of the transfer that increases his net worth, hence his saving, in the year of the transfer.

These results appear to be highly undesirable. They would penalize the property transferors, thereby encouraging retention of property no matter how desirable its transfer might be. And they might often subsidize exorbitant consumption by the transferee, assuming provisions were made for carrying forward any unused deductions for saving. Exception, therefore, appears to be called for to allow property transfer, at death and by bequest (including those to charitable organizations), without the tax consequence in the year of transfer.

Old Age and Survivors Insurance (OASI) System. The proposed tax revisions would materially reduce the costs of saving relative to consumption by subjecting both uses of income to much more nearly equal proportionate burdens. In this new tax environment the occasion for a compulsory, government-funded retirement system, whose adverse effects on private saving and capital formation are receiving increasingly wide notice, should be critically reexamined.

Retirement annuities under the present social security system are financed by payroll taxes collected from those presently employed and from their employers. At no point in the transfer of income from payroll taxpayers to social security beneficiaries are the payroll taxes added to the flow of saving and used to finance real capital formation. Unless beneficiaries save from their retirement benefits amounts at least equal to the reduced saving of payroll taxpayers, total saving must decrease.[8]

Under the proposed income tax treatment of saving, the occasion for relying on a government-administered intergenerational income transfer system to provide retirement income would be greatly reduced. Even if it were deemed necessary to require some form of saving for the provision of retirement income, it would be far better to require such saving through private pension plans or individual retirement accounts. Such saving would be fully subject to the proposed treatment for all private saving; it would, moreover, flow directly into the capital markets for allocation among alternative capital formation uses. The additional capital thereby provided would be the source of the retirement income of the participants; such income, including the return of the participants' contributions, would be fully taxable just as would be the gross returns on all other saving.

Revenue Yield of the Proposed Tax Structure. A major question about the proposed tax structure concerns the size of its tax base, the tax rate or rates applied to it, and the revenue it would generate compared with the present system of taxes. To answer the question it is necessary to estimate the increase in total production capacity and in total income that would be realized compared with the respective amounts under the existing tax system. The requirements for and difficulties in making such estimates have already been noted.

The tax base of the proposed tax structure may be defined, in terms of the national income accounts, as the sum of the excess of the gross returns to capital[v] less capital outlays (private capital outlays plus government deficits or minus government surpluses) plus personal outlays (personal consumption expenditures plus interest paid by consumers to business plus net personal transfer payments to foreigners). Had the proposed tax been fully implemented, its base in 1975, before any adjustments to the tax change occurred, would have been about $1.2 trillion. This tax base would replace the separate tax bases for the federal individual and corporate income taxes, estate and gift taxes, and payroll taxes. In 1975 these taxes raised about $262.5 billion, as measured in the national income accounts.

To raise the total revenues provided by all these taxes under the proposed income tax would have required an average effective tax rate of 22.3 percent in 1975. Had there been no OASI payroll taxes, that is, had a private system for providing for retirement income replaced the existing government system, the required average effective tax rate would have been 14.3 percent.

Burden of the Proposed Tax. The proposed tax has been variously labeled an expenditure tax, a consumption tax, or a consumption-type income tax, implying that its burden would fall solely on consumption, since by hypothesis

---

[v]Gross returns to capital for purposes of this rough calculation are equal to national income less compensation of employees plus capital consumption allowances.

saving would be deductible from the tax base. The proposed tax base, however, would substantially exceed consumption since only *net* saving would be deductible from the tax base.[w] The tax base would include not only consumption outlays but the excess of the gross returns to capital over current capital outlays as well. For the year 1975, for example, the gross returns to capital were $440.2 billion and the excess of the gross returns over current capital outlays (including those to finance the government sector deficit) was $180.2 billion. Thus over 40 percent of the gross returns to capital would be subject to tax. The tax would obviously not be a consumption tax, as it is generally characterized, but an equiproportionate tax on saving and consumption uses of income.

Tax Simplification. Despite these problems and other difficulties that might well arise, the proposed tax holds great promise for simplifying the federal tax system. Since capital gains and losses would not be separate elements in taxable income, the complexities in the present law associated with the tax treatment of capital gains and losses would disappear. Moreover the elimination of capital recovery allowances would contribute substantially to simplification as would the elimination of the corporate income tax and the estate and gift taxes. Most of the provisions now considered loopholes or tax shelters would simply disappear under the proposed tax. Virtually the only source of any such loopholes would be whatever nonbusiness deductions or exemptions were deemed appropriate adjustments to the tax base.

The requirement for maintaining net worth accounts, it is often claimed, would greatly enlarge compliance burdens. For most taxpayers, however, the limited extent and composition of asset holdings and of debts would pose few difficulties. For taxpayers with large and diversified portfolios and debts, the tax accounting problems imposed by the requirement for showing changes in net worth would be little different from those they now face for determining capital gains and losses.

Equity. The proposed tax would eliminate the sources of many of the so-called horizontal equity problems in the present law. Apart from the possible exceptions just noted, the tax base would be substantially broader than under present law with far fewer specific or differential provisions reducing the amount of income subject to tax. Among individual taxpayers with equal amounts of gross income, the difference in their taxable incomes and hence tax liabilities would depend principally on the differences in their costs, including saving, of producing their gross income flows. Other things being equal, the difference in tax liabilities between two individuals with the same amount of gross income would depend on the difference in their saving behavior; the individual saving more would pay a lower tax.

---

[w]This follows from the fact that with the deductibility of saving, no capital recovery allowances of any sort would be allowed, that is, the gross returns on the saving would be included in taxable income.

This difference in tax liability would be inequitable only by reference to tax criteria that require the tax system to burden saving more heavily than consumption uses of income. If the appropriate horizontal equity criterion were deemed to call for equal consumption uses of equal amounts of income, then a tax system of the present configuration moves far along this path. Surely this perception of horizontal equity is no more appealing than its vertical equity counterpart calling for equalization of consumption irrespective of income. And surely it is fair to levy no more tax on an individual who reserves a given part of his income to enlarge the economy's and his own future income than on one who reserves a smaller amount of an equal current income.

Tax Changes to Redistribute the Stock of Capital

These basic tax revisions would materially reduce the existing tax bias against saving. Under the proposed tax the cost of both saving and consumption uses of income would be greater than in a taxless world, but the tax-induced differential in cost would be far less than under the existing tax regime. The eventual response to this restructuring of the tax system would surely be a larger stock of capital and a greater flow of income than might be expected with taxes of the present configuration. But this enlargement of the stock per se has no necessary implication for the shape of the distribution of income. To achieve a more nearly equal distribution of income it is necessary to change the distribution of the ownership of the capital.

This objective might be pursued in the context of the present tax system. Conceivably the existing overall differential tax burden on saving might be maintained or even increased for individuals who are now in the upper income brackets, while special abatements of that excessive tax load might be provided for low-income persons. Existing loopholes might be closed for upper bracket individuals and widened for those in the lower brackets. At the extreme, capital income items might be excluded entirely from the taxable incomes of persons with adjusted gross incomes below some designated amount while taxed at confiscatory rates when received by individuals whose incomes exceed some specified upper amount.

The complexities such arrangements would add to an already impossibly complicated tax system need not be detailed. The inducements they would afford for avoidance and evasion would probably exceed anything previously experienced. But the most fundamental objection is that penalizing saving by anyone, no matter how affluent, is socially undesirable; irrespective of who owns it, a given additional amount of capital adds approximately twice as much to labor income as to capital income.

The appropriate objective is to increase the poor's share of an enlarged stock of capital, when the optimum total stock of capital is determined in an institutional environment that treats saving and consumption of income far more nearly equally than at present. This view urges the tax revisions proposed in the preceding section with such modifications or additional provisions as might be necessary to reduce the present extra cost of saving for the poor in greater proportion than for the rich.

One method for accomplishing this latter objective would be to substitute a tax credit for the deduction for current saving. The current tax base would include saving, as under present law, but a credit based on the amount of this saving would be allowed against tax liability. The credit would be graduated downward; the percent of current saving allowed as a credit against the tax would decrease as income (as newly defined but including saving) increased. For the very poor, for example, the credit might be allowed at a rate of 100 percent, that is, the full amount of their saving might be deducted from their tax liability, with a current refund of any excess of the credit over the tax before credit. The downward graduation of the credit rate could be specified so as to produce any desired differential between poor and rich in the percentage reduction in the cost of saving. Providing any such differential is likely to result in subsidizing the saving of the poor, that is, in reducing the relative cost of their saving below that in a taxless world, if neutral tax treatment is to be afforded saving by the rich.[x]

[x]Suppose, for example, that it were desired to allow the rich a credit precisely equivalent to the deduction for their saving (one that provided income tax neutrality with respect to the effect of the tax on the costs of saving and consumption uses of income) while providing the poor a significantly greater percentage reduction in the cost of future income relative to current consumption. For example, suppose the poor were to receive a 100 percent credit for their saving. In this event their cost per dollar of future income C_p would be

$$C_p = \frac{1}{y}\left[\frac{1}{(1-t)^2}\right] - \frac{1}{y} \qquad (1\text{-}6)$$

where t is the flat tax rate applicable to all income and y is the yield per dollar of saving. The cost per dollar of current consumption C_c would be

$$C_c = 1(1-t). \qquad (1\text{-}7)$$

The relative cost of future income $C_{p_{y/c}}$ would be

$$C_{p_{y/c}} = \frac{1}{y}\left[\frac{1}{(1-t)^2} - 1\right] \div \frac{1}{1-t} = \frac{1}{y}\left[\frac{1}{1-t} - (1-t)\right] \qquad (1\text{-}8)$$

In a taxless world the relative cost per dollar of future income is $1/y$ which exceeds the relative cost per dollar of future income for the poor under the proposed by $[1/y]$ $[2 - 1/(1-t) - t]$. For the rich, for whom a neutral tax credit is to be provided, the cost per dollar of future income C_R must be

$$C_R = \frac{1}{y}\left[\frac{1}{1-t}\right] \qquad (1\text{-}9)$$

Since before the tax credit their cost per dollar of future income would be $C_R = [1/y]$ $[1/(1-t)^2]$, the required credit must be $t/(1-t)^2$. For the rich the cost per dollar of

How one views any such subsidy should depend on one's preferences about redistributing income in the interests of achieving a more nearly equal distribution and the ultimate objective sought by such distribution. If that objective is to provide for more nearly equal consumption, these radical tax revisions are pointless. Since fiscal policymakers come equipped with no particular insights about the consumption preferences of rich and poor, equalizing consumption among them must rely on arbitrary, elitist decisions about how much more and less of what the poor and rich are to consume. For this purpose the principal function of a redistributive tax structure is to finance a highly discriminatory expenditure system. On the other hand, if the objective of income redistribution is merely to serve some ethical or aesthetic desideratum that holds income equality to be good and lovely without (1) overtly constraining individual choices and (2) reducing total income, then I believe the proposed tax changes offer substantial promise for success. Not all the poor would respond equally to the subsidy for saving afforded by the new tax structure; neither would all the rich respond equally to the new, more nearly neutral tax treatment of their saving-consumption choices. But it is surely to be expected that the proposed change in the federal tax system would result in greater total wealth of which the poor would hold a larger proportion; by the same token they would produce and claim a larger share of total income.

Conclusions

I have been a long way around the barn to arrive at suggestions for drastic revisions in the federal tax structure. These revisions have two objectives. The first is to allow us all to be wealthier by providing a tax system that less distorts the choices we all face between saving and consumption of our incomes. The second is to reduce the disparity in the distribution of our incomes by reducing the presently excessive cost of saving relatively more for the poor than for the rich.

The first of these objectives, I think, must engage everyone's support, unless one finds virtue in being poorer than we need be. The second objective, on the other hand, rests on a shaky foundation—the premise that income equality, unlike all other personal attributes, is intrinsically good and lovely while income inequality is inherently bad and ugly. The basic tax revisions I have proposed would, I am convinced, well serve the first objective. The modifications of the

---

current consumption would be the same as for the poor, shown in (1-7). The relative cost per dollar of future income for the rich would then be

$$c_{R_{y/c}} = \frac{1}{y}\left[\frac{1}{1-t}\right] \div \frac{1}{1-t} = \frac{1}{y} \qquad (1\text{-}10)$$

new tax structure that I have suggested would reduce income inequality, but whether they are desirable depends on views for which rigorous analytical support is yet to be found.

There is, however, a pragmatic consideration that strongly urges the modifications, no matter how ambiguous the abstractions on which they are predicated. Egalitarians are in full cry rather than in disorderly retreat. If the wholesome objective of tax revision to reduce the bias against private saving and capital formation is to be accepted, it must embrace a companion objective to which egalitarians assign a higher priority—enhancing the tax bias in favor of the poor. The standard tax reform proposals to serve the latter objective would do so, if at all, only at the expense of increasing rather than reducing the excessive tax burden on saving. An alternative approach is needed. I have presented one such, not with great expectations for its widespread endorsement, but in the immodest hope that it will contribute to a clearer perception of relevant issues and to a more constructive focus in tax policy debates.

Notes

1. Henry C. Simons, whose work on the ideal concept of taxable income has enormously influenced contemporary thought, avowed that his definition of income was explicitly geared to the use of the income tax to equalize economic status. He also avowed that implementation of his proposals in this connection might well transfer the saving function to government, a result he found acceptable if necessary to effect progression in the tax system. See his *Personal Income Taxation* (Chicago: University of Chicago Press, 1938), pp. 25-30.

2. Arthur M. Okun, *Equality and Efficiency; The Big Tradeoff* (Washington, D.C.: Brookings Institution, 1975), p. 3.

3. Ibid., p. 47.

4. C.W. Cobb and Paul H. Douglas, "A Theory of Production," *American Economic Review* Supplement 18 (1928):139-165.

5. There have been numerous examinations of income distribution and its change or lack thereof over time, but all are flawed by frailties of data or concept. Edward C. Budd made a detailed study of the distributions of money income for the years 1947-1968; his study shows no change over time in the degree of inequality. See his "Postwar Changes in the Size Distribution of Income in the U.S.," *American Economic Review* 60 (1970):247-260.

6. See Seymour Fiekowsky. "The Impact of Taxation on Mineral Capital: The Case of Oil and Gas," in *Economics of the Mineral Industries*, edited by William A. Vogely (New York: American Institute of Mining, Metallurgical, and Petroleum Engineers, 1976), pp. 673-682.

7. For an analysis of the effects of the operation of the Old Age and Survivors Insurance System on private saving see Norman B. Ture with Barbara

A. Fields, *The Future of Private Pension Plans* (Washington, D.C.: American Enterprise Institute, 1976), Appendix A.

8. For an analysis of these adverse effects on private savings, see Ture and Fields, *The Future of Private Pension Plans;* a somewhat different analytical approach to the same effect is provided by Martin Feldstein, "Social Security, Induced Retirement, and Aggregate Capital Accumulation," *Journal of Political Economy* 82 (1974):905-926.

Commentaries

2 Commentary:
Michael J. Graetz

In his wide-ranging discussion of the tax system and his proposals for revision Norman Ture touches on each of the major current issues of tax policy: (1) What is the meaning of vertical equity in taxation? (2) What are the effects of the tax system on capital formation? (3) How does one choose between income taxes and consumption taxes? (4) What are the uses and the limitations of the tax expenditure concept? (5) How might the tax laws be simplified? Even though it was not his mission to address systematically each of these questions, Mr. Ture provides helpful observations with respect to many of them.

However, with respect to the focus of the paper—taxation and the distribution of income—Ture's analysis is disappointing. Many questions relating to the effects of taxes on the distribution of wealth are within the economist's capacity to inform lawyers (or, perhaps more importantly, politicians). Unfortunately, Mr. Ture's paper largely fails to do so. One suspects that Ture's real concern is not with the distribution of income, wealth, or consumption, but rather with the displacement of private savings by government expenditures. This paper, however, never explicitly addresses this important issue, but attempts to justify reduced taxation of savings on distributional grounds. In this effort Mr. Ture is not convincing.

The paper is divided into three main sections: (1) ethical-aesthetic arguments concerning distribution of income or wealth, (2) an economic analysis of the "feasibility" of redistribution and the effects on aggregate savings of the current tax system, and (3) a proposal for a consumption tax to replace federal individual and corporate income taxes, estate and gift taxes, and payroll taxes. I will comment separately on each section of Mr. Ture's paper.

Ethical-Aesthetic Arguments

Mr. Ture's effort to address what, following Henry Simons,[1] he characterizes as the ethical-aesthetic issue of redistribution of income and wealth and thus of "the equity criteria of tax policy" is merely the most recent in a long line of generally unsuccessful inquiries. Plato's assertion that no one in society should be more than four times as wealthy as the poorest member probably commands as much (or as little) support as John Rawls's difference principle.[2] The critics Blum and Kalven in their classic examination, "The Uneasy Case for Progressive Taxation,"[3] have enjoyed more widespread acceptance, but one cannot avoid suspecting that the cases for proportional taxation or regressive taxation are equally uneasy. Nevertheless, Secretary of the Treasury Simon could still assert with confidence: "There appears to be a widespread consensus that an element of progression is desirable in the tax structure."[4] This consensus seems related to the fact that for a fixed amount of revenues and pattern of government expenditures and institutional arrangements, progressive taxation will produce after-tax income, wealth, and consumption levels that are distributed less unequally than would proportional or regressive taxation. The persistence of a consensus for progressivity suggests that Henry Simons captured a widespread sentiment when he noted that inequality is unlovely.

But how progressive a tax should be depends on such matters as one's view of the appropriate social welfare function (assuming it is meaningful to talk about such a thing), how one takes into account differences in abilities and tastes, whether one considers an individual's income to be an entitlement or rather a product of the society, and a host of other factors that serve only to verify differences of opinion. Norman Ture is certainly correct when he concludes that "adverting to ethical and aesthetic considerations dictates no unique pattern of income distribution and specifies no particular degree of progression in tax liabilities nor indeed any given configuration of the tax system." Recognizing this, he does not strive to develop any systematic case for a particular distribution of tax burdens. Instead he attempts to reduce the issue to a conflict between "the libertarians and the egalitarians" and to demonstrate that economic equality is "ethically and aesthetically distasteful."

While the contours of our differences are not yet precisely clear to me, I am certain that Norman Ture and I genuinely disagree over ethics-aesthetics. Ture seems to consider a market distribution to be a "just" or "equitable" distribution, but I find it difficult to justify a market distribution as an ethical matter. For me its justification rests principally on grounds of efficiency and consumer sovereignty—a belief that a market economy avoids waste and increases the standard of living even for those with a lesser distributional share. My view of individual liberty includes a determination to preserve the opportunity of consumers to choose to pay enormous sums of money to hear the Captain and Tenille sing "Muskrat Love," regardless of how offended I might be by their choice. These two concerns—for output and consumer sovereignty—require inequality of distribution, and thus equality of result becomes, as Ture suggests, unlovely. Nevertheless, a number of considerations predispose me toward a redistributional policy that reduces inequalities.

First, given our total product, providing some minimum level of welfare seems essential.

Second, a severely unequal distribution has certain qualities of a public good; a failure to revise the market distribution necessarily produces externalities that Mr. Ture ignores. What are the "demoralization costs" of substantial differences of consumption and wealth? Absent significant government efforts to redistribute, at least to alleviate poverty, enforcing the market distribution may prove very costly. Crime and strikes by those charged with law enforcement must affect output. What are the potential effects on output of a minority's withdrawal of cooperation with the background institutions of society? What degree of inequality will produce such a reaction or otherwise inhibit enforcement? Economic productivity depends on certain economic, social, legal, and political foundations. The vulnerability of these institutional arrangements to dissatisfaction with distributional outcomes certainly warrants attention.

Third, one's attitude toward redistributional policy depends on judgments regarding certain background institutions of the economy and government. Whenever certain institutional arrangements such as equality of opportunity or a free market without legal barriers to entry are absent, market distributional outcomes become difficult to defend. In addition the existence of background institutions, for example a body of laws and law enforcement mechanisms generally effective in maintaining a market distribution of property rights, suggests that some portion of output should be characterized as joint or societal rather than individual. In Ture's terms one might ask what portion of marginal productivity is attributable to society's institutional arrangements. Rather than assuming the market distribution to be just, perhaps we should view an individual's distributional entitlements as limited to what he could earn without cooperation and society's protections. Any excess could then be viewed as a return to society and perhaps be divided equally, or unequally given unanimity, or, following Rawls, unequally if this benefits the least advantaged members of the society. At a minimum, using a state of anarchy as a base for comparison may well affect one's judgment about distributional results. James Buchanan, for example, has defended Rawls's difference principle using this approach.[5]

Buchanan's analysis relys heavily on the methodology of the contractarian model and the original position in an effort to persuade or reinforce one's intuitions about distributional issues. William Klein has used the same approach in an effort to justify the proposition that "a fair system is one which distributes material rewards equally among people who are willing to work equally hard."[6] The utilization of the original position contractarian methodology as a justificatory device reflects an effort to develop an analytical framework for addressing the distributional issues that both Ture and I have treated to this point as ethics-aesthetics. In this role the original position has the virtue of requiring one to set forth any formal requirements considered necessary for justice (Rawls, for example, considers generality, publicity, and finality essential) and provides a model for making the complex distributional question more manageable. In assessing its analytic persuasiveness, one must be alert to the underlying biases that make this methodology compelling to Rawls as a justification of outcomes. While this question has been addressed extensively elsewhere,[7] it is worth noting the observation, principally by Ronald Dworkin, that the justificatory power of the original position methodology coupled with a unanimity requirement stems from a notion that the right of each individual to equal concern and respect must be enforced.[8] The appeal of the original position contractarian methodology is undoubtedly due to a more widespread notion of egalitarian rights than Ture seems to acknowledge. Equality of result in this context is eschewed essentially because of its effect on total output and because of the existence of background institutions and formal conditions of "procedural justice" that tend to legitimize results of contractarian or consensual processes.

Ture's paper does not seem particularly helpful in illuminating these genuine but difficult issues. Rather than explicitly addressing considerations such as those that argue for redistribution, he emphasizes the difficulties of implementing a redistributional policy. For example, Ture correctly stresses the serious empirical and theoretical difficulties of ascertaining the distributional consequences of a tax change. To know the distributive consequences of an income tax break for child care expenses, for example, one must estimate the distribution of child care payments by income class. Data about the income levels of those who make payments for child care may not be readily available. Often when such data are available, information concerning the distribution of such benefits is by income classes that are different from those of the tax rate schedule, and such estimates are calculated using a measure of income not known in the tax law. Moreover it is necessary to estimate the effects of the tax change on relative prices and factor incomes to determine the distributional consequences of the change.[9] Typically such difficulties are assumed away. Persons who are nominally affected by a change in the income tax are assumed to bear its entire impact; induced changes on investment activity, consumption, or other economic activity are ignored.

The need for more accurate estimation of the distributional consequences of tax changes is amplified by the existence of considerable tactical posturing in discussions of vertical equity. When the proponent of a change suspects that it will improve vertical equity (as he sees it), whether he argues for the change in terms of vertical equity may depend on whether his views and his predictions concerning the impact of the change correspond to those of the persons he is trying to convince. The numerous proposals designed to adjust the distribution of the income tax burden by changing some provision other than the nominal rate schedule offer some evidence that this is the case.

An additional difficulty in implementing a redistributional policy relates to the dynamic nature of distributional considerations in taxation. A distributional policy considered appropriate for one time may be rejected at a different time. Ture implies that one of the important variables that might affect judgments about distribution is the level of output, but he limits his consideration of this point to a comparison of an "impoverished society" and an "affluent society." But less dramatic differences in output might affect distributional policy. If total output is growing, distributional policy that focuses on the allocation of gains from trade, including producers' and consumers' surpluses, may be appropriate.

In a constant or declining economy, the initial distribution of wealth may be more important. In a subsequent section of his paper Ture seems to assume that even though output changes, a distribution that is proportionally the same as a prior distribution is somehow justifiable, but he does not indicate why the same percentage slice of a different total product is an appropriate distribution.

A society's distributional policy also varies depending on the amount of government expenditures that do not have distributional goals. The larger the level of government expenditures, the higher the rate of taxation, even with proportional rates. In such circumstances it will be difficult to find a consensus for raising tax rates to redistribute income, consumption, or wealth. An increasing level of taxation magnifies the inhibiting effects of taxation on output and inspires persons to substitute untaxed forms of utility. While Ture seems concerned with the overall level of government expenditures, and particularly with its effect in displacing private savings, he never explicitly addresses the relationship between the level of taxation and distributional considerations.

Perhaps the most serious conceptual difficulty in implementing redistributional policy is in identifying the basis for determining equality. The ostensibly simple notion that vertical equity requires unequal taxation of persons in different circumstances is fraught with theoretical difficulties. The tax literature is filled with disputes about whether similar or different circumstances are being compared.[10] For example, if two people have the same income but one has only earned income and the other has only unearned income, are these similar or different cases? Ture's paper raises the fundamental question whether redistributional policy should be concerned with reducing inequality of consumption rather than of income or wealth. This important issue bears particularly on the choice between consumption and income taxes. Ture argues that equalizing consumption is necessarily the goal of "income-equalizing tax policy." He asserts that consumption is all that can be equalized because income will necessarily become unequal if there are any differences in individuals' consumption-savings patterns. Ture concludes that the objective of equalizing consumption "if explicitly identified, would be rejected out of hand." Ture's comments in this regard are not persuasive. By treating absolute equality as a goal, he creates a straw man that he then proceeds to destroy. The essential question is whether after-tax income, consumption, or wealth should be more equal or less equal than now, given the necessary trade-off in terms of decreased output for reducing such differences. If one concludes that any move toward equality is desired, it is then relevant to ask whether it is income, consumption, or wealth that should be made less unequal.[11]

Discussions of redistributional policy have tended to assume that reducing inequalities of income or wealth is the appropriate goal. Income is typically selected on the grounds that the power to consume, or consumption opportunities, rather than actual consumption, should be made more nearly equal. But if one views savings as nothing more than deferred consumption, this position seems difficult to maintain, because, as Ture demonstrates, income taxation will produce different tax burdens depending solely on when an individual chooses to consume. Such differences seem arbitrary unless one views an individual's savings as something other than deferred consumption, perhaps as according status or social or political power. But even if one views savings as conferring status or social or political power, why should redistributional policies focus only on the status or power represented by savings? We would generally reject efforts to redistribute other forms of status (for example, status from being beautiful or from being a doctor or even a professor) and have made little effort to neutralize political power (for example, of congressmen, regulators, or bureaucrats). The major recent effort to reduce inequalities of political power has, in fact, concentrated on the disparities of such power produced by differences in the ability to make (and in tastes for) political contributions. At a minimum, a significant amount of savings should be required before its status or power justifies redistribution.

While it is not possible to develop fully such a case in this commentary, I believe that a convincing case can be made for focusing redistributional policies on consumption rather than income, and I suspect that resistance to this idea would likely result from concern with differences in inherited wealth, differences in an individual's initial endowments. This analysis therefore would focus attention on differences in lifetime consumption and on accumulated savings that remain unspent at death. Under a consumption model one might view deathtime transmission of property to other persons as consumption and therefore an appropriate subject of redistributional policy. Alternatively one might consider differences in wealth that result from gifts or inheritances to be morally arbitrary and thus a separate appropriate object of redistributional policies.[12] Ture advances additional arguments for increasing the share of capital of those at the bottom of the scale. By so identifying the goals of redistributional policy, one can make an independent distributional case for consumption taxes, either with separate accession or estate taxes or by including the transmission of property in the consumption tax base. If the redistributional goal is to reduce differences in consumption, a progressive tax on consumption should address this goal with less deadweight loss than an income tax. A case for consumption taxation may thus rest on redistributional grounds.

Economic Analysis

As a lawyer with little exposure to aggregate economic theory, I should confess at the outset that this section of Ture's paper is difficult for me to evaluate. Ture seems to set for redistribution policy an impossible goal to achieve a more nearly equal distribution "without making society as a whole poorer." The paper gives little attention to the question of the various effects on total output of different schemes of taxation.[13]

Mr. Ture provides an example that illustrates one possible effect of income taxes on output and provides some description of the varying effects of federal income and estate taxes and state income and property taxes on savings, but he does not supply adequate information to be helpful in illuminating the choice among various taxes on efficiency and distributional grounds. For example, Mr. Ture argues the inhibiting effects on savings of taxes on wealth and income. In addition he indicates the dampening effect on savings of the social security system which is financed exclusively from payroll taxes. Any tax, it seems, will produce flows from the private sector to finance current government expenditures and will thereby decrease private savings. Thus one is concerned with the differential effects of various financing schemes, an issue that the paper does not address.

In addition I consider Ture's emphasis on pretax incomes misplaced in evaluating redistributional effects of taxes. A judgment about the efficacy of redistributional policy necessarily turns on the distribution of after-tax income, consumption, or wealth, but Ture does not present evidence bearing on this point. His concession that "redistributive tax and expenditure structures *may* reduce inequality in the distribution of after-tax income" is not illuminating.

The crux of Ture's economic argument is that the amount of capital should be increased, that such an increase will expand total output, and that an increase in the capital-labor ratio will not increase the share of total output that flows to owners of capital. Mr. Ture asserts that "changes in the aggregate capital-labor ratio result in changes in total output and income and in the payments per unit of capital and labor service but do not change the shape of the distribution of income." I am willing to assume the validity of Ture's use of neoclassical theory as a basis for his analysis,[14] but his conclusions seem to be based on certain unstated assumptions. For example, Mr. Ture states: "[T]he Lord devised the specific relationships between the amount of the change in production input and the change in total output. An important property of this production function is that the percentage change in output is the same as the percentage change in the input, holding the other inputs constant." It seems that the proportional change in output that results from a percentage change in input would vary with the absolute level of inputs assumed. If Mr. Ture is assuming a particular mix of inputs from which he is measuring the change, he does not specify what the fixed input point is. On the other hand, if he is assuming a constant elasticity of output with respect to inputs regardless of the level of inputs, he must have in mind a particular production function, presumably a Cobb-Douglas production function.[15] Although he refers to Cobb and Douglas' work, he does not indicate whether his analysis requires one to assume a Cobb-Douglas production function.

Likewise Ture's conclusion that a change in the ratio of capital to labor will not affect the pretax share of output that flows to capital or labor seems to be sensitive to his assumed production function, in particular to the elasticity of substitution between capital and labor. For capital and labor to receive the same proportionate share of output as capital is increased relative to labor, the relationship of wages to profit rates must change in the same proportions. That is, if capital doubles relative to labor, profit rates must halve relative to wages. The elasticity of substitution between capital and labor must be equal to one. If the elasticity of substitution between these inputs is greater than one, capital's share of output increases as the ratio of capital to labor increases. An elasticity of substitution equal to one is an attribute of a Cobb-Douglas production function,[16] but a recent empirical study estimates that the elasticity of substitution may fall between 1.148 and 1.245.[17] While I am not competent to judge whether Ture's assumptions are reasonable, I am troubled by his failure to make them explicit and must wonder what effect different, but equally reasonable, assumptions might have on his analysis.

Tax Revision

Mr. Ture proposes a consumption tax to replace the federal individual and corporate income taxes, estate and gift taxes, and payroll taxes. He provides some suggestion of the necessary "average effective rate" to produce equivalent revenue, but he gives no suggestion of the marginal rate schedule he would propose. Thus it is impossible to know the distribution of the tax burden that would result from his proposal. If he intends the distribution of this tax to fall in the same way as the taxes he would replace, he should so indicate. One suspects that such a distribution would require quite high marginal rates on consumption, probably over 100 percent, at least for individuals who now have adjusted gross incomes of $100,000 or more. If he intends a distribution of the tax burden that tends to increase or decrease inequality in the after-tax distribution of income, consumption, or wealth compared to the existing system, he should make his rate schedule explicit and attempt to estimate the gains or losses in terms of total output. Without such information it is impossible to evaluate the equity of his proposed system.

Mr. Ture claims that a consumption tax would likely be simpler than current law. This would probably not be the case if, as Ture suggests, net worth accounts were required of taxpayers. However, William Andrews has presented a detailed argument that a cash-flow consumption tax would eliminate many of the complexities of the income tax.[18] But many problems would remain. Distinguishing consumption from production expenses would continue to be difficult and depending on the schedule and the goals of the consumption tax might well be exacerbated. In addition, distinguishing borrowing for consumption from business and investment borrowings would likely be required. It would also be necessary, for example, to determine whether education expenses, medical expenses, charitable contributions, and state income taxes are consumption expenditures. Until a statutory scheme for a specific consumption tax is detailed, it is impossible to know whether the tax would be simpler or more complex than present law. The transitional problems associated with moving from the present system to a consumption tax would undoubtedly prove enormously complex. Ture does not comment on this problem, and Andrews treats it only briefly.

Ture's consumption tax proposal does explicitly provide a criterion against which income tax provisions might be evaluated. Ture's analysis and comments "in defense of income tax loopholes" nicely illustrate the different conclusions one reaches depending on whether one is using a consumption ideal or, say, the Haig-Simons definition of income as a model. As Ture suggests, one's attitude toward capital allowances such as depreciation and other deductions for savings, such as for individual retirement accounts, is critically dependent on the criterion for evaluation. Unlike many commentators and, as he notes, unlike the tax expenditure budget, Ture makes his frame of reference explicit. Ture acknowledges the existence of many provisions that reduce the tax burden on savings—for example the tax exemption for state and local bond interest and the tax advantages accorded individual retirement accounts and qualified pension plans—but he makes no effort to evaluate the distributional consequences of these provisions, their impact on the overall quantity of or returns to savings, or their effect on total output.

Conclusion

Mr. Ture and I disagree over the ethics-aesthetics of redistribution policy; he does not seem to share my predisposition toward reducing inequalities. His economic arguments do not convince me that redistribution is not feasible, although I recognize dangers in minimizing the difficulties of implementation. I do share his preference for a consumption-type tax over income taxes, but I suspect that each of us would recommend rate schedules that would vary in important respects. I would also insist that a consumption tax proposal be coupled with substantial estate or accession taxes. I hope that this conference will either demonstrate that our differences are not as great as they seem or illuminate why our differences have not produced a greater divergence in the policy recommendations we might offer.

Notes

1. Henry C. Simons, *Personal Income Taxation* (Chicago: University of Chicago Press, 1938), pp. 18-19.

2. Rawls's difference principle provides that "social and economic inequalities are to be arranged so that they are both (a) to the greatest benefit of the least advantaged and (b) attached to offices and positions open to all under

conditions of fair equality of opportunity." John Rawls, *A Theory of Justice* (Cambridge: Harvard University Press, 1971), p. 83.

3. Walker Blum and Harry Kalven, Jr., *The Uneasy Case for Progressive Taxation* (Chicago: University of Chicago Press, 1963).

4. Treasury Document 76-25. Treasury Secretary William E. Simon to Mrs. Llewellyn Lowe of Silver Spring, Md., quoted in *Tax Notes,* December 27, 1976, p. 11.

5. See James M. Buchanan, "A Hobbesian Interpretation of the Rawlsian Difference Principle," *Kyklos* 29 (1976):5-25. Buchanan states:

[T]he interpretation is more readily understood if we assume that cooperative action necessarily introduces a dramatic shift in the technology of producing income or product. The jointness aspects of the basic structure of social arrangements become predominant. By way of a simplified economic illustration, we might say that the Rawlsian model allows Crusoe and Friday to commence fishing with a boat once agreement is reached, whereas in anarchy this degree of cooperation is not possible, and each man has to fish without a boat. The cooperative arrangement involves participation in the provision and use of a genuinely public good. In this framework, it becomes impossible to impute separate income shares to the two parties, Crusoe and Friday, since the whole production is clearly a joint product.

Crusoe and Friday agree to act jointly, to become partners in social arrangements; gross production increases dramatically, but there is no means of imputing separate shares (p. 8).

Buchanan then suggests that unequal divisions will be acceptable only to the extent that the least advantaged person is better off with equal sharing. If work incentives are such that unequal sharing will make both persons better off than with equal sharing, this will be agreed to; but otherwise, the least advantaged person will withdraw cooperation and force the system back to equal sharing.

6. William Klein, *Policy Analysis of the Federal Income Tax* (Mineola, N.Y.: Foundation Press, 1976), pp. 26-30.

7. See, for instance, Norman Daniels, ed., *Reading Rawls: Critical Studies on Rawls' A Theory of Justice* (New York: Basic Books, 1975).

8. See Ronald Dworkin, "The Original Position," in *Reading Rawls,* edited by Norman Daniels, pp. 16-52. See also, for instance, T.M. Scanlon, "Rawls' Theory of Justice," in *Reading Rawls,* pp. 171-179.

9. Michael J. Graetz, "Assessing the Distributional Effects of Income Tax Revision: Some Lessons from Incidence Analysis," *Journal of Legal Studies* 4 (1975):351.

10. For excellent treatments of this subject, see Louis Eisenstein, *The Ideologies of Taxation* (New York: Ronald Press, 1961), p. 160; Walter Blum and Harry Kalven, *The Anatomy of Justice in Taxation,* Occasional Papers (Chicago: University of Chicago Law School, 1973), pp. 23-30; Martin Feldstein,

"On the Theory of Tax Reform," *Journal of Public Economics* 6 (1976):77; Richard A. Musgrave, "ET, OT, and SBT," *Journal of Public Economics* 6 (1976):3.

11. This is the essence of the debate between Alvin C. Warren, Jr., and William D. Andrews. Compare Warren's "Fairness and a Consumption-Type or Cash Flow Personal Income Tax," *Harvard Law Review* 88 (1975):931, 934-936; to Andrews, "Fairness and the Personal Income Tax: A Reply to Professor Warren," *Harvard Law Review* 88 (1975):947, 949-952. See also Andrews, "A Consumption-Type or Cash Flow Personal Income Tax," *Harvard Law Review* 87 (1974):1113.

12. There is, of course some conflict between any tax on wealth and tax neutrality between present and future consumption, but as Professor Andrews indicates, this does not necessarily imply a preference for income taxes over consumption taxes. See ibid., p. 955.

13. In this respect Ture's paper may be contrasted with recent efforts by economists to measure the efficiency-equity trade-off. Many of these papers have tended to focus on the "optimal progressivity" of an income tax. For a description of this literature, see David Bradford and Harvey Rosen, "The Optimal Taxation of Commodities and Income," *American Economic Review* 66 (1976):94-101.

14. This stems from my inability to judge the force of the so-called Cambridge criticism in this context. See, for example, Charles E. Ferguson, *The Neoclassical Theory of Production and Distribution* (London: Cambridge University Press, 1969), pp. 254-258.

15. Ibid., p. 76.

16. R.G.D. Allen, *Macro-Economic Theory: A Mathematical Treatment* (New York: St. Martin's, 1967), p. 51.

17. Ernst Berndt, "Reconciling Alternative Estimates of the Elasticity of Substitution," *Review of Economics and Statistics* 58 (1976):59, 65.

18. William D. Andrews, "A Consumption-Type or Cash Flow Personal Income Tax," *Harvard Law Review* 87 (1974):1113.

3 Commentary:
Martin Feldstein

Norman Ture has given us a broad and provocative paper. To keep my task manageable I will limit myself to two subjects: his discussion of the ethics of redistributive taxation and his central analytic argument that taxes cannot raise the disposable income of the poor. I will begin as Ture did with the ethical basis of redistributive taxation.

I agree with Ture's most basic point that the *logical* foundation for the ethics of income redistribution is shaky. Perhaps that is the reason that I did not find Ture's own discussion of the ethics of income redistribution very useful. I believe that professional philosophers have a comparative advantage vis-à-vis economists in the development of ethical principles. A Mill can create a case for utilitarianism, a Rawls for the maximin principle, and a Nozick for the idea of just entitlement. I find each such system enlightening but none of them utterly compelling. In the end I believe that economics—and the theory of taxation in particular—cannot hope to develop convincing ethical principles but must concentrate on drawing out the implications of alternative ethical standards.

Stated somewhat differently, I do not believe that as economic scientists we can say what form of taxation is optimal. All we can do is say: If your criterion is X, the best tax rule is Y. Conditional statements of this sort do not constitute a trivial problem, as a great deal of recent work on the theory of optimal taxation demonstrates.[1]

The ethical principle that underlies most of the formal and informal analysis of optimal taxation has been utilitarianism. A specific form of this principle implies that if all men had identical tastes and supplied fixed amounts of effort,[a] taxes should be set to make all consumption equal. Since economists do not believe that work effort is fixed, they reject the conclusion of equalizing consumption in favor of tax schedules that balance redistribution and the excess burden of distorting labor supply.

Not everyone accepts the appropriateness of utilitarianism as an ethical standard.[b] A related but distinct criterion of just taxation would be: Levy taxes so that each individual's *sacrifice* of utility is the same.[2] Note that this principle can support a progressive tax but not a redistributive tax. If the principle is accepted, the economic problem becomes how to design a tax schedule to

---
[a]I shall ignore the issue of saving at this point. As Ture notes, differences in tastes for leisure and goods are sufficient for his argument.

[b]Note I am dealing with the logically prior issue of the appropriate standard rather than the feasibility of implementation.

achieve the required equal sacrifice when individuals' labor supply responds to taxation.

It is clear that Ture rejects both the utilitarian and equal-sacrifice principles. Even with no distortion and no excess burden, he sees nothing desirable about equalizing utility or equalizing utility sacrifice. There is nothing wrong with Ture's view. It is, as he says, a matter of taste on which individuals are free to differ. I assume Ture's ethical principle is close to Nozick's: Individuals are entitled to their property, and no redistributive tax is justified. That still leaves open the question of how to finance the necessary services of government. I would be interested in knowing more about Ture's view of the correct principle to follow in answering that question.

In a few places Ture becomes carried by the momentum of his own arguments to conclusions that he would recognize do not follow and are not correct. Suppose we grant for the sake of argument his premise that "equalization of economic status and economic freedom, therefore, are incompatible." Neither that premise nor any of his earlier arguments then supports the conclusion: "An institutional thrust *toward* economic equality is necessarily equivalent to the *elimination* of the free market mechanism" (emphasis added). Surely society can take the smallest step toward greater economic equality without thereby eliminating the free market. Even if some degree of equalization is incompatible with a free market and even if every bit of equalization reduces economic freedom, it is not true that a small amount of redistribution totally eliminates the free market—unless, of course, markets are only to be regarded as free if there is no interference at all, an interpretation that would make Ture's statement a tautology.

Although most of Ture's second section deals with the ethical foundation of progressive taxation, Ture cannot resist the opportunity to criticize some of the political rhetoric of tax reform. Unfortunately he tends to use the shallowness of these rhetorical statements as implicit evidence that all statements about tax reform are equally incorrect. Thus in a single paragraph Ture correctly objects to the slogan that "money earned by money should be taxed at least as much as money earned by labor" but then goes on to argue that all redistributive fiscal policies represent the results of rhetoric that deludes the public to accept "arrangements that frustrate our efforts to distinguish, to differentiate, to follow our own stars."

We can argue about the just ethical principles for taxation—utilitarianism, maximin, entitlement—but we should not believe that as economists we have any particular skill in making such arguments. We will be more useful if we separate the attempt to convince others about fundamental principles from the more mundane but more productive task of drawing out the implications for tax policy of different ethical principles and different descriptions of economic behavior.[3]

This brings me to Ture's discussion of the feasibility of redistributive taxation. The central point of Ture's paper is that such fiscal policies cannot raise the income of the poor: "Any reduction in after-tax income inequality is likely to be attained at the expense of everyone's being poorer in absolute terms; most people will be little, if at all, better off in an absolute sense as a result of the redistribution of disposable income." This conclusion is based on the logically possible notion that high taxes on capital income can reduce the capital-labor ratio, thereby lowering the marginal product of labor and the wage rate. What workers gain by receiving government transfers may be more than offset by the fall in wages.

Such an argument about the *possible* shifting of a capital income tax is of course logically correct and widely recognized in the public finance literature.[4] Indeed, a few years ago I examined the more general problem of the optimal redistributive tax in an economy with two factors of production, with the supplies of those factors responding to changes in their net returns, and with those net returns determined jointly by a Cobb-Douglas production process.

But while Ture's "perverse" result is logically possible, there is no reason to believe that it is empirically correct. The total income of labor will fall only if the reduction in capital is very substantial. Ture presents no evidence to suggest that capital accumulation is so responsive. It is wrong to assert (as Ture does) that taxing capital income necessarily reduces private capital accumulation. Moreover, even if a tax on capital income does reduce saving, the magnitude of the reduction is likely to be too small to cause the perverse result that Ture posits. More specifically, in the appendix to this chapter I use a model similar to Ture's to show that introducing a small tax on capital income and transferring the proceeds to labor *always* raises the total disposable income per worker.

It is clear, however, that there is a limit to the extent to which taxes can be used to redistribute income from capital to labor. The law of diminishing returns applies to taxation as well as to everything else. The higher the initial tax rate, the less likely it is that a further increase can succeed in raising the net income of labor. At some point short of a 100 percent tax rate, Ture's argument is likely to be true for the empirically relevant savings elasticity. But my analysis shows that in the simplified model economy that Ture has in mind, an increase in the tax rate from an initial value of 50 percent will raise the disposable income of workers unless the elasticity of savings with respect to the net rate of interest exceeds two-thirds. While such a strong response is certainly not impossible, all the evidence suggests that it is not true. Michael Boskin's recently estimated elasticities of 0.3 to 0.4 are "substantially larger than virtually all previous estimates."[5]

Even if private saving were quite sensitive to the net rate of return, there would be no reason to reach Ture's conclusion that fiscal redistribution is impossible. If the government taxes capital income but saves all the proceeds (for example, by reducing the national debt), the capital stock will actually increase and wages will rise.

The inflow of foreign capital would also prevent wages from falling. Recall that any reduction in the capital stock would not only lower wages but also raise the pretax return on capital. This would attract foreign capital to the United States. If the United States tax paid on the capital income earned by foreign investors in the United States can be fully offset against their tax liabilities at home, the free international flow of capital would keep the capital-labor ratio unchanged in the United States.[c]

Until this point I have gone along with Ture's focus on the taxation of *capital* income. I believe, however, that Ture overemphasizes capital income. After all, there is substantial inequality in the distribution of income derived from professional, managerial, and other personal services. The issue of redistributive taxation would not disappear if all capital income were exempted from tax. What happens to Ture's argument about the perverse effect of redistributive taxation if we focus on the high tax rates levied on the personal service earnings of those with high earnings?

Following Ture's line of argument, we would expect that a higher tax on what I shall refer to as high-skill labor earnings will depress their supply, that is, induce the taxed individuals to consume more leisure and fewer goods.[d] How would the reduced supply of labor by these high-skilled individuals affect the wages paid to low-skilled individuals? To state this question in different words, are high-skilled individuals substitutes for low-skilled individuals in production, or are high-skilled individuals complements in much the same way that machines are? If high-skilled labor is a substitute for low-skilled labor, a tax-induced reduction in the supply of high-skilled labor will actually raise the wage rate of the low-skilled. The general equilibrium effect of the tax would thus reinforce the direct redistributive effect rather than reduce it. Unfortunately we still do not have the empirical evidence needed to answer this question.

Although I have emphasized my reservations about Ture's analysis, I take the general message of his analysis to be correct: Discussion of redistributive taxation should look at the general equilibrium effects of the taxes that are levied. To assess the net effects that taxes do have or that proposed taxes could have, we need to know more about behavioral and technological parameters than we currently do. The econometric evidence that is beginning to be produced will provide a better basis for discussions like this one.

---

[c]It might even be argued as a rough approximation that without capital mobility, wages would still be maintained through international trade in commodities by what is known as the factor price equalization mechanism.

[d]This does not necessarily occur. The high-skilled workers may work harder to recoup part of the fall in their income that results from the higher tax rate.

Appendix 3A

I have postponed to this appendix two technical issues that I want to present but do not want to impose as a barrier to noneconomists in reading the text of my comments.

I will deal first with the issue of the relation between the taxation of capital income and the net rate of saving. Since so many policy discussions assert that a higher net yield would induce more private saving, it is worthwhile to examine why this may not be true. Consider for example a person at age 40 who wishes to save for retirement at age 65. This individual expects to obtain an average real after-tax return of 4 percent. For every $100 per year that he saves during the next 25 years, he will be able to dissave $357 per year from the time he is 65 until he is 80. In light of this opportunity he decides to save $1400 per year and thus have $5000 of dissaving each year when he is retired.[e]

Now consider what happens if his net rate of return rises from 4 percent to 5 percent. This implies that every $100 per year saved from age 40 through 64 will buy substantially more retirement consumption; more specifically, with a 5 percent return the individual could dissave $422 per year instead of the $357 obtained at 4 percent. Faced with this lower price of retirement consumption, the individual is very likely to increase the level of planned retirement consumption. *But if the new level of retirement consumption does not increase sufficiently, current saving will actually fall.* For example, if the individual decides to increase his annual retirement dissaving from $5000 to $5500, he can actually lower his current saving from $1400 to $1303. Only if annual retirement dissaving increased to at least $5900 would current saving increase.[f] Since we do not know how the demand for retirement consumption responds to its net cost, it is not possible to say whether saving would increase or decrease in response to a higher net rate of return.

---

[e]This description of individual choice is of course a stylized representation of the actual process of saving. Some kind of calculation of this sort must implicitly be made if a change in the rate of return is to have any effect.

[f]Economists will recognize that two separate issues are involved here. First, a reduction in the price of retirement consumption (an increase in the net rate of return) will increase the level of retirement consumption if the *un*compensated demand for retirement consumption is a decreasing function of its price. Theory requires only that the compensated demand be a decreasing function of its price. Nevertheless, a nonnegative marginal propensity to save implies that the uncompensated demand will also be a decreasing function of price.

Second, a reduction in the price of retirement consumption will increase the rate of saving only if the demand for retirement income has a price elasticity absolutely greater than one. By definition, saving is equal to the product of retirement consumption and its price. The elasticity of saving with respect to that price is thus equal to one plus the elasticity of retirement consumption with respect to its price. If this negative elasticity has an absolute value less than one, it follows that the elasticity of savings with respect to this price is positive. And since this price varies inversely with the net rate of return, in this case the elasticity of savings with respect to the net rate of return is negative.

Although this example was developed in terms of retirement saving, similar ambiguity applies to the other and less important motives for saving. We simply do not know how a rise in the net rate of return would affect the amounts that individuals save for such things as children's education, the future purchase of a home or of consumer durables, or to have funds available for emergencies.

I turn finally to the question, When does taxing capital income and transferring the proceeds to labor raise net labor income? (Readers will recognize that this slightly simplifies Ture's analysis by ignoring the small amount of capital owned by workers.)

In general notation we write the Cobb-Douglas technology $y = k^a$ where y is national income per workers, k is capital per worker, and $a = 1/3$ is the elasticity of y with respect to k. This implies a wage per worker of $w = (1-a)k^a$ and a gross rate of return to capital of $r = ak^{a-1}$. If capital income is taxed at rate t, the "grant" g that can be paid to each worker is $g = trk$. Labor income per worker is $W = w + g$. By substitution we obtain

$$W = (1 - a + ta)k^a \tag{3A-1}$$

Let us assume with Ture that the supply of capital depends positively on the net of tax rate of return $r_n = (1-t)r$; more specifically

$$k = C_0 r_n^E \tag{3A-2}$$

where C_0 is a constant and $E > 0$. Since $r_n = (1-t)r$ and $r = ak^{a-1}$, we have

$$k = C_0 [(1-t)\,ak^{a-1}]^E \tag{3A-3}$$

or

$$k = C_1 (1-t)^{E/[1+E(1-a)]} \tag{3A-4}$$

We can therefore write labor income as

$$W = C_1 (1 - a + ta)(1-t)^{aE/[1+E(1-a)]} \tag{3A-5}$$

The problem to be solved can be stated as: Under what condition is $dW/dt > 0$? Rewrite (3A-5) as

$$W = C_1 (1 - a + ta)\, e^{b \ln (1-t)} \tag{3A-6}$$

where $b = aE/[1 + E(1-a)]$, and note

$$\frac{\partial W}{\partial t} = C_1 e^{b \ln (1-t)} \left\{ a - \frac{(1-a+ta)b}{1-t} \right\} \qquad (3\text{A-}7)$$

Thus $\partial W/\partial t > 0$ whenever

$$a > \frac{(1-a+ta)b}{1-t} \qquad (3\text{A-}8)$$

or, substituting again for b we get

$$a > \frac{(1-a+ta)aE}{(1-t)[1+E(1-a)]} \qquad (3\text{A-}9)$$

or

$$(1-t)[1+E(1-a)] > (1-a+ta)E \qquad (3\text{A-}10)$$

Note that at $t = 0$ equation (3A-10) implies

$$1 + E(1-a) > (1-a)E \qquad (3\text{A-}11)$$

a condition that is always satisfied. Thus a new small tax can always succeed in redistributing income to labor.

We can rearrange the terms in (3A-10) to obtain

$$(1-t) > E[1 - a + ta - 1 + t + a - at] \qquad (3\text{A-}12)$$

or

$$(1-t)/t > E \qquad (3\text{A-}13)$$

Thus as long as the elasticity of the capital stock with respect to the net rate of return is less than $(1-t)/t$, an increase in the capital income tax rate will raise the net income of labor if labor receives the tax proceeds. Even at $t = 0.5$, this is satisfied for any $E < 1$.

It is more natural to think about the elasticity of the savings rate with respect to r_n. Note that in equilibrium growth

$$\sigma y = nk \qquad (3\text{A-}14)$$

where σ is the saving rate and n is the growth of the effective labor force. Thus

$$\sigma k^a = nk \qquad (3\text{A-}15)$$

or

$$k = [\sigma/n]^{1/(1-a)} \qquad (3\text{A-}16)$$

Since $k = C_0 r_n^E$, it follows that

$$\sigma = C_2 r_n^{(1-a)E} \qquad (3\text{A-}17)$$

The condition that $E < (1-t)/t$ is therefore equivalent to the condition that the elasticity of savings with respect to the net rate of interest be less than $(1-a)(1-t)/t$. Thus for $a = 1/3$ we have the conclusion stated in the text that even with a tax rate of 50 percent a higher tax will raise the net income of labor if the savings elasticity is less than two-thirds.

Notes

1. Readers who would like to know more about this body of research should consult the very useful articles by Agnar Sandmo, "Optimal Taxation: An Introduction to the Literature," *Journal of Public Economics* 6 (1976):37-54, and David F. Bradford and Harvey S. Rosen, "The Optimal Taxation of Commodities and Income," *American Economic Review, Papers and Proceedings* 66 (1976):94-101.

2. For a discussion of this view, see Richard A. Musgrave, *The Theory of Public Finance: A Study in Public Economy* (New York: McGraw-Hill, 1959).

3. My views about this are developed more fully in Martin Feldstein, "On the Theory of Tax Reform," *Journal of Public Economics* 6 (1976):77-104, especially in section 2.

4. See, for example, Michael J. Boskin, "Taxation, Saving and the Rate of Interest, Working Paper No. 135 (New York: National Bureau of Economic Research, 1976); Peter A. Diamond, "Incidence of an Interest Income Tax," *Journal of Economic Theory* 2 (1970):211-224; Martin Feldstein, "Incidence of a Capital Income Tax in a Growing Economy with Variable Savings Rates," *Review of Economic Studies* 41 (1974):505-513.

5. Ture's case could be strengthened by dropping the assumption of a Cobb-Douglas technology. With a low enough elasticity of substitution between capital and labor, even a small fall in the capital stock can have drastic effects on wage rates. See Feldstein, "Incidence of a Capital Income Tax," and Boskin, "Taxation, Saving and the Rate of Interest."

4 Commentary:
Boris I. Bittker

Mr. Ture has given us an essay worthy of extended discussion. I hope to unravel only a few of the threads in the sombre drapery that he has prepared for the progressive income tax. (Whether the drapery is to cover its grave or to prepare the tax for a new life on earth is not wholly clear.) Mr. Ture in his paper is by turns an ethical philosopher, political theorist, Biblical exegete, logician, cultural anthropologist, economist, libertarian, sociologist, econometrician, gadfly, modelmaker, tax reformer, politician, and hired gun.[1]

Perhaps the last two labels require a word of explanation. I call Mr. Ture a politician because, despite his distaste for egalitarianism, he evidently regards it as the wave of the future and concludes that if you can't beat it, you should join it. (I am reminded of students who tell me that they are revolutionaries at heart, but plan to work within The System.) As for the "hired gun" metaphor, before you take exception to it I hasten to explain that it is often used to define the lawyer's trade. It seems apt here because Mr. Ture, who regards the progressive income tax as outrageously biased against the rich, offers to help society increase this bias with a tax credit for savings that will vanish as the taxpayer's income rises. What could be more lawyerlike than this willingness to give the client whatever he or she wants by representing the client "zealously within the law," as required by canon 7 of our code of professional responsibility?

The many roles exemplified by Mr. Ture's essay challenge the commentator, who, even if unpersuaded or stubbornly unrepentant, must admire this multidisciplinary product of a one-man think tank. Indeed, a biographical note informs us that Mr. Ture is a corporation, evidently of the one-man variety. I mention this fact only because it suggests that one of the points I would make about the corporate income tax is already well known to Mr. Ture, namely that it can be an opportunity rather than a burden. In fact some of my best friends approach it in the mood of the gourmet who dislikes Swiss cheese: discarding the solid matter to get at the holes.

Mr. Ture is in the mainstream of economic thought, however, in disregarding the curious fact, well known to lawyers, that enterprises that could easily be conducted as proprietorships or partnerships deliberately seek a corporate charter, displaying no fear of the monstrous antisavings bias that we are told is embodied in the corporate income tax. Nor does Mr. Ture address the question, which I have frequently put to economists without getting much of an answer, whether the corporate income tax insures to the benefit of those remaining entrepreneurs who never pay it because they operate in unincorporated form.

Perhaps this question could be answered by the econometric model envisioned by Mr. Ture.

The Progressive Income Tax

If I understand his essay correctly, Mr. Ture's view of the progressive income tax can be reduced to the following syllogism:

1. The real agenda of those who advocate a progressive income tax, hidden behind "flights of oratory" and "shimmering sequins," is the achievement of "equality of economic status."
2. This objective can be achieved in an affluent and specialized economy only by "homogenization of the society," which requires "elimination of the free-market mechanism," which in turn will produce a drab and all-pervasive "sameness of cultural attributes."
3. While "[t]astes in matters of aesthetics and morality vary" and "do not lend themselves to rigorous analytical determination," the foregoing rigorous analytical determination exposes the advocates of progressive income taxation for what they really are—dangerous, misguided, and naive people like Henry Simons and Arthur Okun.
4. Repent while there is still time!

Let me start with the premise that advocacy of progression cannot be divorced from advocacy of "equality of economic status." Assume for the moment that progression, if applied rigorously, would in fact greatly reduce economic inequality. Mr. Ture, in fact, is evidently of two minds about this proposition. At one point he seems to argue that no matter what is done to them, the rich will rise again, taking strength like Antaeus from each fall; and he vouches the practices of "collectivist societies" (the Soviet Union? Or is this an oblique reference to the United States now that it has fallen to the egalitarian onslaught?) as evidence that affluent societies are ruled by an iron law of unequal wages that cannot be repealed even by the Internal Revenue Code. Despite these intimations that progression will not really reduce economic inequality, Mr. Ture is not willing to take a chance. A more daring libertarian, however, would be willing to let the egalitarians indulge their harmless fancy, knowing that the invisible hand would soon return the economy to its preordained levels.

Let us, therefore, assume at least for argument that a progressive income tax rate will permanently twist the Lorenz curve into a straight line denoting equality.

If progressive income taxation reduces economic inequality, does it follow that advocates of the former ineluctably favor the latter? I would like to rephrase the question: Does the fact that Florida levies a tax on hotel room occupancy prove that its legislators are trying to exclude tourists and will not rest content until they have returned the state to the Seminoles? My point is that one can favor progression—believe that it is a good way to match the citizen's tax bill to his ability to pay[2]—and simultaneously, with complete rationality, consistency, and intellectual honesty, believe that economic equality will be either substantially narrowed or only marginally affected by progression, and that either of these results is desirable or undesirable but preferable to the effects of other ways of financing the government.

I am not sure whether Mr. Ture could stomach the income tax even if it exempted savings and were proportional. If these changes would win him over, however, he would thereby accept a tax that increased the "cost" of labor relative to leisure. Would that prove that he wanted to discourage work and that his secret agenda was a homogenized world of drab beachcombers?

A word now about Mr. Ture's theory that affluent societies, being diverse, specialized, and richly endowed, either depend for their success on, or necessarily produce (or is it both?), highly divergent wage levels. This observation is plausible, and everyone at this conference is a beneficiary of this iron law of unequal wages. But I cannot quite leap from this observable system of rewarding specialized and scarce talents to the conclusion that progressive income taxation has destroyed, is on the verge of destroying, or inevitably will destroy this system. In virtually all institutions of our society—the universities with which we are especially familiar, the federal civil service, and business organizations save at the very top—the salary scale from bottom to top is confined to a ratio of 1 to 10 or thereabouts, namely, $6000 to $60,000, and this range encompasses as varied a set of talents as can be found or even imagined. If income taxation has compressed this range to a dangerous extent, the evidence escapes me. It is obvious, of course, that the cost of living has skyrocketed for the upper middle class, as compared with the days before World War I. But this change is primarily attributable to the disappearance of a class of low-paid domestic servants and to similar forces that Mr. Ture does not propose to roll back.

The income tax has also risen exponentially, but one cannot conduct a land war in Asia on the cheap. Before concluding that the income tax is a form of "expropriation," one must ask whether those who are taxed are getting, on the whole, what they want from the government. In this connection, am I wrong in detecting an infrastructure of philosophical anarchism in the Ture essay? Some citizens objected to taxes that were to pay for bombing Vietnam back to the Stone Age; at times, Mr. Ture seems to have similar objections to taxes used to finance anything at all unless it commands universal voluntary acceptance.

Finally I am puzzled by Mr. Ture's choice of targets. As I read the essay, the income tax's sin of sins is a disruption of the pretax trade-off between savings and consumption. But income taxation also disrupts the pretax trade-off between work and leisure. Why not restore *this* pretax balance by exempting income from personal services and taxing only income from capital? Other taxes, for their part, disrupt the pretax equilibrium between still other alternatives. If the Ture essay explains why one trade-off is more worthy of perpetuation than all others, his rationale escapes me. I suspect that his unarticulated premise is a special reverence for savings, but perhaps I am mistaken.

The Effect of the Current Tax System on Savings

A central theme in Mr. Ture's essay is that the failure to exempt savings from federal income taxation (and from other taxes as well) produces a poorer society, thereby inflicting losses on labor and capital alike. He illustrates this message by the story of a model society that taxed capital income and thereby reduced the returns to both capital and labor.

Though I could find no error in Mr. Ture's arithmetic, my intuition—nagging though, or perhaps because, unscientific—suggests that there might be more to say. By consulting the editor of the *Journal of Heuristic Studies,* I located the society described by Mr. Ture, which proved to be an island near the Garden of Eden.[a] I then asked its prime minister to flesh out Mr. Ture's rather sparse description of his country. The prime minister gave me some supplemental information.

The 10 percent of the population that were big savers before imposition of the capital income tax described by Mr. Ture responded to the tax by reducing their consumption in an effort to restore their pretax level of savings, and, while not fully succeeding, they managed to get fairly close to the status quo ante. The prime minister attributed this to the fact that instant gratification is religiously offensive to these people, who prefer savings banks to bowling alleys. He also mumbled something about the "income effect" of the tax, saying that Professor Musgrave could explain what he meant.

The 90 percent of the population that received the negative income tax payments financed by the tax on capital income did not increase their consumption as much as had been predicted. They proved, in short, to be savers *manqué*—and what they previously *manqué*-d was money. Some of them, in fact, saved more than they received in tax transfer payments, because of a change in personal psychology that the prime minister variously attributed to a "revolution of rising expectations" and to a "critical mass" of funds.

The net result, I was told, was an increase in gross national product, to the point where the government was about to indulge itself in a previously undreamed of luxury—a council of economic advisers. The prime minister added ruefully, however, that the increased penchant for savings had made the society rather drab ("homogenized" was his phrase), because everyone was now studying the theory of compound interest instead of dancing in the streets.

---

[a]A commune of pre-Columbian hippies described by the Bible and made even more alluring by Mr. Ture.

He continued—once he learned that his country's story might be spread before a group of American experts, there was no stopping him—to tell me more. Within a few years there were enough savers in the society to seize control of the government. They then voted to subsidize agriculture and shipping, to regulate some other industries, to impose a tariff, and to imitate the free enterprise systems of western Europe and the New World in many other ways. Some time later (this brings us up to date) a wave of populist sentiment brought the current prime minister into office on a platform that included an increase in the tax on capital gains from 10 to 11 1/2 percent.

All hell broke loose. The country's most distinguished consulting economist said that this would discriminate against savings when compared with the perfect balance prevailing in a no-tax world. The prime minister told me that to evaluate this assertion, he had hired a competing (though less famous) firm of consulting economists, known as Second Best Systems Analysis, Inc., whose corporate logotype consisted of a large question mark, rampant, emblazoned on the Great Seal of the United States Treasury. They told him that it would be impossible to decide whether the proposed tax would move the society closer to or further from the no-tax world that all hands posited as the ideal, without analyzing all other economic programs in the society including taxes, expenditures, tax expenditures, and expenditure taxes,[b] in short, everything. But to do this would be expensive, he was told, since to find out anything about anything, you must first know everything about everything. The prime minister told me that on hearing this he was inclined to rely on his intuition, which told him that the bias against savings in the proposed tax was probably more than offset by the prosavings bias in other governmental programs.

Despite my earnest effort to be accurate, it may be that the society I have just described is not the same one that Mr. Ture had in mind. If so, I offer as my excuse the old Hungarian proverb: "What is heuristic for the goose is heuristic for the gander," which, rendered into the vernacular, means, If a game is worth playing, two can play it.[3]

---

[b]The concept of an expenditure tax is unfortunately less well known than tax expenditure, its justly famous Siamese twin. A tax expenditure, as everyone knows, is a provision in a tax law that exempts someone from a tax imposed on similarly situated persons; hence it is tantamount to a subsidy to the exempted person and can be viewed as an expenditure made through the tax system. An expenditure tax is the converse: a provision in an *expenditure* law that *denies* someone a subsidy that is paid to similarly situated persons; hence, it is tantamount to *taxing* the disqualified person, and should be viewed as a tax effected through the expenditure process.

Both of these undesirable phenomena can be identified only by determining whether the exempted person (in the case of a tax expenditure) or the disqualified person (in the case of an expenditure tax) is treated differently from others who are similarly situated. As respects tax expenditures, this process is proceeding apace in the Treasury, the Office of Budget and Management, the General Accounting Office, the Congressional Budget Committee, and in universities throughout the nation. It is devoutly to be hoped that equal time will be given to ferreting out expenditure taxes, so that they can be properly reflected in the budgetary process and so that jurisdiction over expenditure taxes can be transferred from the various appropriations committees of Congress to the agencies that have legitimate authority over taxes—the House Ways and Means Committee, the Senate Finance Committee, and the Internal Revenue Service.

I report my conversation with the prime minister as a prelude to a description of what I see as an unresolved conflict—producing a fundamental, if not fatal, internal inconsistency—in Mr. Ture's essay. His conclusions are deduced almost exclusively from abstract premises that are unsullied by any but the most tangential association with real life. Thus there are no market imperfections, no regulations, no labor unions, no subsidies, no governmental expenditures, and no capitalized expectations in the world that is suddenly disrupted by the advent of an income tax. The remedies deduced from an inspection of this society, however, are then prescribed for another society (our own), with almost no attention to the latter's etiology. To be sure, we are given a quick survey of the taxes with which the patient is now infected (without mention, however, of sales taxation). Mr. Ture paints a somber picture of an assault on the rich by a band of sans-culottes, led apparently by a few Brookings radicals under the improbable banner of that patron saint of the Chicago Libertarian Pantheon, Henry Simons. The only bright spot was percentage depletion, and even that ark of the covenant has now been defiled. Does our society have no other features—imperfect markets, government regulations, loans, subsidies, tariffs—that affect the savings-consumption ratio?

There is, of course, nothing wrong with brainstorming, but social remedies derived from an intense contemplation of mathematical symbols should not, in my view, be offered too quickly to real-life patients. Moreover, one might also wonder why we should strive to make the actual world conform to an idealized society that operates without compulsion when we know that organized society cannot dispense with compulsion.

Be that as it may, it is surely the case that one cannot ascertain by a process of deductive reasoning from abstract premises whether our society now has more or less actual savings than would be produced by the model society used by Mr. Ture as his touchstone. We are told, for example, that "the tax-induced increase in the cost of future income relative to consumption was about 77.6 percent in 1976," but we are given no estimate of the *government-induced decrease* in the cost of future income relative to consumption through subsidies, regulations, loans, and so forth.

One can be sure, of course, that the actualities of real life have produced a different mix of investments than we would have gotten from an economy that was less influenced (or distorted) by collective social decisions. But I do not think that Mr. Ture is seeking merely to prove that we suffer from inefficiencies in the allocation of a suitable stock of investments, producing, for example, "too much" housing at the cost of "too little" clothing. He is telling us, in addition, that the aggregate is less than it would be in an idealized society; and unless I misinterpret his tone this is his main point. But I find in the essay no factual foundation for this conclusion.

If the absence of facts is to be excused because the necessary econometric model is not yet operational, then it seems to me that the conclusion should be presented, nakedly but without shame, as an intuitive judgment. For myself, I prefer a good shouting match to a contest in which unrealistic premises are employed by one side, or even by both.

A corollary of Ture's methodology is that if it turns out that our society, because of the lamentable intrusion of social decisions, has more savings than it would have under ideal circumstances, he would cheerfully support measures to reduce this swollen stock of investments to the "proper" lower level. I wonder.

My own reaction to all this is distrust of the idealized model presented by Mr. Ture, curiosity about the promised econometric model, and skepticism about measuring social success by reference to a wholly voluntary (and hence wholly disorganized) society that, if it existed anywhere, would lead to lives so solitary, poor, nasty, brutish, and short that even the drab homogenization feared by Mr. Ture would be welcome. I would prefer an incremental view of social decisions, in which the wisdom of moving from one position to another is judged by whatever insights one can bring to bear, human frailties are frankly admitted, and occasional comparisons with a never-never land are recognized as having a marginal value at best.

The Proposed Tax Allowance for Savings

I am not persuaded by Mr. Ture's argument in favor of a tax allowance for savings, but I also cheerfully disclaim any independent expertise regarding its economic wisdom. Let me offer instead a few random observations of a more limited character.

1. The federal income tax now contains a network of explicit tax allowances for savings, most notably the provisions excluding employer contributions to qualified pension and profit-sharing plans from the taxable income of the employee and the related deductions for contributions by the self-employed. Moreover these provisions, which have been called "the biggest tax shelter of all,"[4] exempt the investment income generated by the contributions from taxation during the taxpayer's preretirement years. Thus no income tax is imposed on either the contributions or the investment yield until the benefits are received in cash by the retired taxpayer. For millions of taxpayers this is probably the only form of savings that now occurs, and for them Mr. Ture's proposed credit would institute a dollar-for-dollar reduction of their tax liability to replace the current exclusion and deduction in computing taxable income.

How much of an increase in aggregate savings can be expected from this change? Will the resulting reduction in revenue be recouped by increased taxes on taxpayers above the proposed vanishing point for the credit? Where will the consumption-savings ratio be when equilibrium has been restored?

2. Unless progression and personal exemptions are eliminated from the revised income tax envisioned by Mr. Ture, it will reward families who can shift

their savings to high-bracket individuals and their consumption to low-bracket members of the group, especially because the estate and gift taxes are slated by Mr. Ture for extinction. Example: Parents save $6000 each year for each child and take the resulting deduction. The savings are then transferred, tax-free, to the child, who spends this amount for food, clothing, shelter, and education, deducts the personal exemption, and pays a tax on the balance at a lower rate than the parents.

I can almost hear the quick response from the macroeconomic heights from which economists look down on lawyers: "That's a detail; fix it up; amalgamate the family income." OK, we'll put that on the agenda, just below the plan to tax faculty tuition remission plans. My point: Don't count your simplification chickens before the tax experts remove the barn door.

3. The proposed repeal of the federal estate and gift taxes. But Congress just took two-thirds of the potential taxable estates off the rolls. Not only that, but the deed was done by legislation bearing the title "Tax Reform Act of 1976." And to cap the climax, a vote for that legislation got congressmen on the good-guy list of Ralph Nader's Tax Reform Research Group. Mr. Ture, meet your allies.

4. Assertion: With a rate of 22.3 percent (or, with Old Age and Survivors Insurance payroll taxes eliminated, 14.3 percent), the proposed tax would have raised $262.5 billion in 1975. Comment: In the sage words of Mr. Ture, "initial impact revenue estimates" are "grossly misleading" if they disregard the secondary effects of a tax change and "[t]he one thing that may be confidently said about such estimates is that they are certainly wrong."

5. Here is a test case for the political viability of the proposal, in its "strong" version (without pandering to egalitarian sentimentality by imposing a vanishing feature). Let us suppose that the late J. Paul Getty spent no more than a typical senator for food, shelter, and clothing; but instead of using his entire income for these purposes (as did the hypothetical senator), he was able to save $5 million a year. Who will volunteer to write a speech in favor of equal taxes for these two equal-expenditure individuals and to brief the president in preparation for his first press conference after signing the bill? Speaking for myself, I'd rather be one of the reporters.

6. Justice to the expenditure-tax concept is long overdue, and a little affirmative action would be appreciated. Perhaps the members of this conference would care to go on record in this regard.

7. It looks to me as though this tax proposal, though designed to beat the egalitarians at their own game, has abolished the personal exemptions. But maybe my arithmetic is wrong.

Notes

1. The only role that might have been added, with profit, is constitutional historian. Had he delved into the reports of the United States Supreme Court,

Mr. Ture would have found that *Pollock* v. *Farmers' Loan & Trust Co.*, 158 U.S. 601 (1896), holding that the 1894 federal income tax was unconstitutional, left room for a resuscitated income tax (even before the Sixteenth Amendment was ratified) that would have been close to his heart's desire. The *Pollock* case permitted, *mirabile dictu,* an income tax on income from personal services (whether spent or saved), but *exempted* income derived from property. Didn't someone once say that those who ignore history are doomed to repeat it?

2. I share Mr. Ture's admiration for Blum and Kalven's *The Uneasy Case for Progressive Taxation.* But the title is not *The Fallacious Case for Progressive Taxation* or *The Conclusive Case against Progressive Taxation.* My view is that the case for progressive taxation is no more "uneasy" than the case for proportionality or for preferring one tax base over another (Galvin and Bittker, *The Income Tax: How Progressive Should It Be?* Washington, D.C.: American Enterprise Institute, 1969).

3. See Richard B. Goode, *The Individual Income Tax,* rev. ed. (Washington, D.C.: The Brookings Institution, 1975), p. 57: "Neither deductive reasoning nor the factual evidence now available clearly shows whether the economic differences [the impact on savings, consumption, and labor] among the three tax bases [income, expenditures, and wealth] are great or small."

4. Frederick Hickman, "Pension and Profit-Sharing Plans: The Quintessential Tax Shelter?" Department of Treasury Press Release, no. S-336 December 5, 1973.

5 Commentary:
Richard Musgrave

Ture's provocative paper presents with much fervor a position with which I was expected to disagree. I shall meet this expectation. But I also find that his case in some important respects, is not stated well and fails to reflect much of the literature of recent years. I thus have the double task of clarifying his position as I think it should be stated and then subjecting it to a critical review. In so doing I follow Ture's outline, commenting in turn on certain ethical aspects of distribution, the economic feasibility of changing distribution, and applications to tax reform.

Ethical Underpinnings

I am pleased that Ture begins his paper with a discussion of the ethical underpinnings of the distribution issue. I would have been more pleased, however, if his treatment had reflected the rather extensive and serious discussion that this topic has received in recent years among social philosophers and even economists. A lot has happened in this area since the Blum-Kalven volume to which Ture refers. Economists like to think that rigor ends where the maximizing process of the market ceases to apply, leaving the field to visceral (to use Ture's term) assertion. But such is not the case. Rigorous and careful thinking is not limited to neoclassical economics but can be exercised also on ethical and social problems.

I disagree with Ture's proposition that the egalitarian case is made more easily for low-income than for high-income economies. Indeed I believe the opposite is true. The low-income economy can less afford the efficiency costs that may be involved in redistribution, and it can less afford to forego saving. Nevertheless it is useful to examine the ethical problem in successive stages, beginning with a simple setting and advancing to a more complex one.

Suppose there are three individuals with different earning capacities, but (to begin with) no substitutability between work and leisure and no saving or capital formation. All three work the same hours, but H on the high side catches ten fish a day; M in the middle catches seven, while L on the low end catches but four. The problem of social justice then involves simply the distribution of fish consumption. Should consumption be left as unequal as indicated by the differentials in natural endowments, or should there be some specified degree of equalization?

Simple though this case is, it does pose the central problem. Philosophers like Nozick, following the Lockean tradition, argue that each person is entitled to the fruits of his labor. Justice is defined in terms of how property is acquired, not in terms of the resulting end-state of distribution. Differential consumption is just, provided it is not based on stealing someone else's catch. A social compact will be needed to provide protection against theft, but that is all. Philosophers like Rawls, following the Kantian tradition, view the problem differently. They hold that endowment by birth is arbitrary and carries no ethical sanction. Rather, talents (the total catch of 21) should be viewed as a joint product that should be distributed fairly. It is the end-state that matters and a fair solution (a solution that, in line with the Kantian principle of universality, is independent of one's initial lot) calls for an egalitarian outcome, with seven fish per person. Still another tradition is based on Bentham's utilitarian maxim that it is reasonable for society to maximize total welfare. The just or rational solution then depends on the capacity of various individuals to derive utility from consumption. If this capacity were comparable and uniform—or, better, if it were considered uniform by social judgment—the result would again be egalitarian. If not, other results may ensue.

In noting these alternative approaches, I do not wish to defend one over the other as being ethically correct. I personally prefer the Kantian tradition, but my purpose here is only to show that there can be a careful discussion of alternative models, far from the visceral form to which Ture refers. The choice between them cannot be made without value judgment, and Ture is entitled to consider equalization of consumption abhorrent and to reject it out of hand. But he should have made it very clear that this is his personal view and without claim for general validity.

Nor is there any reason why departures from the Lockean position, be they in the form of partial or total equalization, must interfere with consumer choice between products. Suppose that for each person the catch is divided between 30 percent herrings and 70 percent flounders. There is no reason why, in either the Kantian or Benthamite tradition, the fanciers of herring should not be permitted, after their allotments are received, to engage in trade with those who prefer flounders. It is simply wrong to assert that a move in the egalitarian direction need involve uniformity in consumption.

Moving on to a more realistic setting, I now allow for substitutability between work (consumption of fish) and leisure. This complicates matters in several ways. Since substitution is possible, the transfer from H to L that would be required to reduce inequality now involves an efficiency cost as the marginal rate of substitution between consumption and leisure (the net wage rate) comes to differ from their marginal rate of transformation in production. This cost arises on both the paying (H) and receiving (L) side of the scale. Given this fact, society must face a trade-off between the gains from equalization and the efficiency cost involved. As seen by most economists, this would lead to less than complete equalization, although an extreme social welfare function of the Rawlsian (maximin) type would still retain that result. It is strange that this aspect of efficiency cost, which has been at the center of the economic discussion of optimal taxation, has been overlooked in Ture's paper.

Matters are complicated further if we allow for differences in preferences in the goods-leisure choice. Mr. Jones, a member of the H group, may have a high-leisure preference, while Mr. Smith, also in that group, may have a high-fish preference. Consequently, Mr. Jones will sharply reduce his work effort as the tax is imposed while Smith will not do so or may even increase it. Thus the tax contribution made by Smith will be substantially greater than that of Jones. The redistributive process discriminates in favor of those with high-leisure preference and against those with high-goods preferences.

Ture dislikes such discrimination, and I agree.[1] The just solution would be to tax not on the basis of *actual* consumption (which, since saving has been excluded so far, is identical with actual income) but at its *potential* level. The Kantian requirement would be for people to reveal their potential income, in which case the solution is straight forward. If they do not, a capacity tax is difficult to implement and a second-best solution is needed. The question, then, is whether the loss of justice (in terms of horizontal equity) that arises from the discrimination between Smith and Jones more than outweighs the gain derived from reduced inequality between them and the L group. In my judgment this is not the case. To reject recognition of the distribution issue because preferences differ is to let the tail wag the dog. Damages to horizontal equity should be allowed for, along with efficiency cost when weighing the desirable degree of equalization, but they do not exclude it.

Taking a final step toward realism, I next allow for saving and capital formation. Now a distinction must be drawn between a tax on income and a tax on consumption. As all economists agree, a tax on income (which includes interest in the base) discriminates against saving, whereas a tax on consumption does not. But both approaches involve discrimination between goods and leisure. This being so, there is no a priori efficiency case for the consumption standard. Adding the consumption-saving choice introduces a further dimension, but it does not change the nature of the trade-off problem.

Feasibility of Redistribution

I now turn to the second topic in Ture's paper, whether redistribution is feasible. He concludes, with considerable satisfaction though I believe incorrectly, that economic analysis says no. Directing himself to long-run policy effects, he attempts to derive this result from certain elementary verities: (1) Capital formation raises productivity and hence total output. (2) Capital formation is reduced by the taxation of capital income. Given that (3) capital income weighs more heavily in the higher income ranges, we find that (4) progressive taxation is detrimental to capital formation. Consequently (5) a tax on capital income reduces future national income or earnings. He then notes that (6) both labor and capital share in income growth. Postulating (7) a Cobb-Douglas production function so that the reduction in income leaves factor shares unchanged, he concludes that (8) redistribution involving a tax on capital income with transfer of the proceeds to low-level labor income will be to everybody's advantage. More specifically, it will (a) reduce total income, (b) increase income at the lower end of the scale by less than the transfer, and (c) reduce income at the upper end of the scale by more than the tax. Given the Cobb-Douglas function, these conclusions are correct, but they do *not* show that the policy has failed to change the pattern of distribution. As Ture's own illustration quite correctly records, the overall level of future income has fallen, but the low-income share in *net* income has risen. Given his Cobb-Douglas assumption, the latter follows quite independent of the fact that the level of future income has declined. But Ture's illustration goes further and shows that the *absolute* level of labor's net income has declined. This result is needed to "prove" that everyone loses, but it is arbitrary and follows from Ture's particular assumptions with regard to the elasticity of capital and labor supply. These elasticities, taken at 1.5 for saving and 2 for labor, are on the high side. If they are less, as they are likely to be, it is quite possible for the absolute level of labor's future net income to increase. Assuming, for instance, that labor supply is inelastic while retaining the assumption with regard to capital, we find that the net income of the lower group in Ture's illustration would rise to $511. Moreover many different results might be arrived at if other forms of the production function were used.

Ture's fixation on changes in future income only misinterprets the problem. Not only is it questionable whether *all* will suffer a decline in future income. Measuring the social cost of the policy only in terms of reduction in future income also ignores the increase (due to the initial decline in saving) over the earlier level of consumption. While the resulting shift in the timing of consumption is not irrelevant—it poses a problem in intergenerational equity—the decline in future earnings is not to be confused with the efficiency cost of redistribution. This cost, rather, consists of the deadweight loss that arises because of the interference in the efficient choice between present and future consumption. A similar cost would result from a policy that raised the rate of saving (and hence the level of future income) above its optimal level. All this has been pointed out at length in recent literature, and fairly sophisticated attempts have been made to estimate the cost involved. The cost thus defined is substantially less than suggested by Ture's inappropriate concept of redistribution cost. However, I should note (on behalf of Ture's position) that the efficiency cost in terms of saving effects is but part of the problem. The efficiency cost, in terms of resulting changes in work effort, may be more severe, especially so since imposition of a negative income tax involves high marginal rates of tax on earnings at the lower end of the scale.

In short, progressive tax-transfer policies to change the pattern of distribution *are* possible, but they are undertaken at the price of an efficiency cost (deadweight loss) with regard to both saving and leisure. Moreover the result is a shift in the timing of consumption to the disadvantage of the future generation, and interferences with horizontal equity may raise. How one values these effects compared to those of reduced inequality is a matter of judgment, but clearly the result of reduced inequality can be obtained by the use of such policies. In my scale of values, some efficiency cost (which in the end does not turn out to be so large after all) is worth the outcome; and the ensuing shift in consumption toward the present does not worry me greatly, given the outlook for continuing increases in per capita income due to both technological progress and a declining rate of population growth. Moreover, why should just the initial generation of low-income people, whose consumption would in fact rise due to the transfers, be asked to forego this benefit and identify with the interests of low-income people in the future generation? Class solidarity across generations may be a bit much to demand.

Implications for Tax Reform

I now turn to the implications of all this for the future direction of tax reform.

The Antisavings Bias

Ture is correct in noting that the tax system reduces the net rate of return to saving (I prefer this to his terminology of "increasing the cost of saving") and thus tends to reduce the rate of capital formation. The income tax reduces the return on capital income, and imposition of an additional corporation tax further reduces it. So does the property tax. On the other hand the effective rate on capital income is lightened by such factors as faster-than-economic depreciation and tax exemption or preferential treatment of capital gains. Public services that increase the return to capital should also be noted. Nevertheless there remains a substantial net tax on the return to savings. Based on national income data, Ture estimates the effective rate on the gross return to saving at 44 percent, increasing the cost of capital by 77 percent in his preferred terminology.

This burden on capital income tends to be mitigated by features in the tax system that, in his words, protect capital income against "the full fury of the presently punitive tax system." These features, therefore, are all to the good from Ture's point of view, although he might have made more allowance for efficiency costs that arise because such benefits are given only to particular forms of capital income. Given my perspective of broad-based income tax reform, the very same features appear as unfortunate loopholes.[a]

To be sure, a loophole (or, similarly, a tax expenditure) can be defined only in relation to what is considered the proper tax base. Thus my view of the present capital gains treatment as a loophole is based on the premise that the intention is to impose an income tax of the Simons-accretion type. If instead the consumption base were taken as a norm, the loophole concept would be applied to items of consumption that are excluded and not to income sources that are omitted. In some cases, such as imputed rent, the two types of loophole overlap, but this is the exception only. The question then is which of the two bases we want to have.

Income Tax versus Consumption Tax

If one's sole concern were with the efficiency cost that arises from distortions in the present-future consumption choice, the obvious solution would be to omit capital income from the income tax base, thus taxing wage income only, or to replace the income tax altogether by a tax on consumption.[b] But do we really wish to do this? Such a shift still leaves discrimination between goods and leisure so that there is no clear preference on efficiency grounds. Moreover the choice of tax base is not just an issue of efficiency. It must be viewed also in equity terms. Otherwise there should be ready agreement that head taxes are the ideal form of taxation.

To draw a fair comparison, consumption taxes of the sales or value-added tax type should be distinguished from consumption taxes framed in the form of a personalized, progressive expenditure tax, as suggested by Irving Fisher and

---

[a]While I classify myself with this group, let me note the following. If income tax loopholes are defined as tax provisions that permit A to pay less than called for by his full income position, then there must also be negative loopholes—situations of excess taxation. I have always held the view that the corporation tax is a case in point and that equitable treatment of corporate-source income calls for full imputation to the owner, but no additional corporation tax. I thus favor full integration. At the same time, I do not support the case for partial integration, which, though relieving dividend income, would retain the use of corporate retention as a tax shelter. Moreover, serious difficulties would arise in the treatment of foreign investment income.

[b]The base of a tax on wage income and a tax on consumption income, as viewed in ex post national income accounting terms are the same only if capital income = investment + net exports + government expenditures − indirect taxes. There is no reason why this need be the case. Nor will an individual taxpayer be indifferent between the two bases. The wage base will be preferred by someone who has accumulated in the past, thus receiving interest income and expecting to consume in the future; and the consumption base will be preferred by someone who has not accumulated in the past, thus receiving wage income and expecting to save in the future. A taxpayer will be indifferent between the two options only in the hypothetical case in which the choice is presented to him at the very beginning of his economic career and assuming that bequests and gifts made are counted as consumption by the donor.

Nicholas Kaldor. My reference here is to the latter type only, since otherwise the income base (because of its association with a personal-tax approach) would be the clear winner. Comparing the two tax bases in this spirit, which is to be preferred? Which base comes closer to reflecting a fair measure of tax-paying capacity or index by which people can be classified as being in the same position and hence as deserving of the same tax? My present view is that equal position had best be defined in terms of equal options. Defining equal options further in terms of equal present value of potential future consumption, I conclude that under very stringent assumptions, consumption plus gifts and bequests (*not*, I emphasize, consumption alone) is the proper base. However, these stringent assumptions do not apply in practice. Looking at the matter from a more practical point of view, I continue to view income as the more equitable base, and I am willing to pay some efficiency costs in buying this advantage.

But I do not wish to be too dogmatic on this point. I would not object to a program that would (1) replace part of the income tax yield, say one-third, by real progressive and personal expenditure (consumption plus bequests or gifts made) tax, while (2) assuring a complete and full accretion base for the remaining two-thirds that is derived from the income tax. What I do object to—and this is precisely what I detect Ture's approach to suggest—is that one should dismantle progressive income taxation under the banner of consumption-tax reasoning, thereby further extending preferences to capital income, without also imposing real progressive consumption tax.[c] This, I believe, will not only prove a roundabout and somewhat hidden way toward reduced progressivity, but (worse) lead to an unprincipled view of our major personal tax, where no barriers remain from which the tax base can be protected against the onslaught of special-privilege seekers. As I have argued repeatedly in discussions with Professor Bittker and others, I recognize that in detailed application the concept of an idealized base (be it in terms of income or consumption) runs into difficulty, but this does *not* void the need for establishing a norm from which a consistent tax-base policy can be derived.

Ture's Tax Base

A good deal has been written, from Fisher over Kaldor to Andrews, about how the consumption tax base should be defined, with general agreement that consumption should be defined as initial cash + money income + net borrowing + net gifts and bequests received + net sale of assets, − final cash balances. The tax base thus defined equals consumption. Strangely, Ture does not follow this accepted pattern. Instead he proposes to define the base as consumption

---

[c]In the same context, I beg advocates of a consumption tax to call it by its true name (consumption tax) and not to muddle semantics, which is never a harmless matter, by referring to it as a consumption or "cash-flow type income tax."

plus the gross return to capital minus gross investment, thus exceeding the consumption base by the excess of gross capital income over investment. I find this difficult to follow. If we wish to arrive at the base by deducting saving from income, we have

$$C + I + N + G - T - CCA - S = B$$

where C is consumption, I is investment, N is net exports, G is government purchases, T is tax revenue, CCA is capital consumption allowances, S is household saving plus retained earnings, and B is the tax base. But given the accounting identity

$$I + N - CCA - S - D = 0$$

where $D = G - T$ is the deficit, we arrive at the conclusion that $B = C$.[d] Thus the tax base, similar to that derived by Kaldor, equals but does not exceed consumption.

Growth Incentives under the Income Tax

Since Ture notes (though regrets) that egalitarians are not yet in full retreat, he concedes that future tax reform cannot avoid making a bow in that direction. Since he holds that the only way to redistribute income is to redistribute the capital stock, he proposes a system that subsidizes low-income saving. While I do not accept his reasoning, I have some sympathy with his prescription. In fact the approach is rather similar to a suggestion of my own made in 1963 when I examined the problem of growth with equity—long before it became a popular topic—and urged that ways be found to break the link between measures to encourage an increased rate of growth and the tendency for appropriate tax changes to reduce the progressivity of the tax structure.[2]

I then noted that preferential treatment might be given to capital income received by low-income groups. This still seems a good idea, although I wonder whether it will be effective, simply because the rate of tax applicable to low-income taxpayers does not give sufficient leeway by way of tax exemption to make a significant difference. To be effective, not only preferential tax treatment but a positive subsidy may be needed. Strangely enough, we continue to do the opposite, as reflected in the measly return to savings bonds. If one really means business with regard to redistribution of wealth, other and more powerful approaches (say severe death duties or profit participation schemes) should be considered. A further alternative is to substitute public for private saving which, of course, does not mean that there must be corresponding public investment.

How Much Growth?

Ture's paper proceeds on the premise that increased saving and investment are necessarily good things. More accurately one should say that depressing investment below its optimal level (as well as raising it above that level) involves an efficiency cost. If the optimal level is defined as that at which the pretax rate of return equals the investor's rate of substitution between present and future consumption, then as explained by Feldstein and others the level of investment that prevails under the present income tax regime is considered suboptimal. Hence an increase would be called for on these grounds.

But there are some difficulties with this conclusion. First, the approach implies the perfect functioning of a competitive macro system, including capital markets, as the result of which the market rate of interest is given an optimality sanction similar to that which might be bestowed on prices in a competitive product market. Yet our system swings between unemployment and inflation and calls for stabilization policies. Moreover a choice must be made between policies (fiscal monetary packages) that will have quite different implications for the rate of interest. Given this situation I find it somewhat difficult to apply the concept of *the* efficient rate of saving or of growth. Second, the economy may be in a position where the Keynesian lesson still holds, that is, where an increase in ex ante saving may raise unemployment rather than reduce investment. Surely few at this point would wish to urge measures to greatly increase the rate of saving, at least not without also adding others that would provide additional investment incentives.

Foreign Investment

A further consequence results from Ture's emphasis on the gains that labor stands to reap from increased saving and investment. Given this emphasis, a sharp distinction should be drawn between domestic and foreign investment. My point is not so much that part of the returns to foreign investment accrue to a foreign (rather than the United States) treasury, but that the gains of increased capital formation to labor will in this case be reaped not by United States but by foreign labor. Thus there results an amusing joining of interest between domestic labor and foreign capital and vice versa. Given this situation, and given the fact that some 20 percent of United States' saving flows into foreign investment, one of the major prescriptions in Ture's policy ought to be a discontinuation of the tax preferences now extended to foreign investment; indeed he should prescribe measures to limit foreign investment to a level below that which would result from neutral treatment.

Conclusions

Ture's proposition that egalitarian objectives and economic freedom are incompatible makes good campaign rhetoric, but I see little merit to it.

Clearly a tax transfer system that would result in complete equalization, thereby breaking any linkage between effort and reward, is incompatible with a market-directed allocation of resources. Clearly allocation of resources by a planning process would have to be substituted. Such a process may well be less efficient than is the market, although the latter has its problems as well. More important, as I see it, allocation of resources by a planning process (including assignment of labor services to jobs) would call for a degree of concentration of political power that might impose serious threats to human as well as economic freedom.

But the bogeyman of total equalization is not, or at least should not be, the issue; and the serious concerns that complete equalization would raise do not follow for a policy of moderation, including moderate equalization via a progressive tax-transfer scheme. On the contrary, Karl Marx (and so Ture, in viewing the point through the other side of the window) was wrong when, in the *Communist Manifesto,* he urged the use of progressive income taxation as a first step toward the downfall of the system. My understanding of social dynamics teaches me the opposite lesson. Serious concern with poverty and abatement of inequality (apart from being pleasing to my sense of social justice) renders a positive contribution to the viability of a decentralized and private-property based economic system.

Notes

1. Richard A. Musgrave, "Maximin, Uncertainty, and the Leisure Trade-Off," *Quarterly Journal of Economics* 88 (1974):625-633.
2. Richard A. Musgrave, "Growth with Equity," *American Economic Review* 53 (1963):323-333.

Discussion

Philosophical-Ethical Issues

Bittker: Is Mr. Ture discussing redistribution as an end in itself? For many supporters of a particular kind of tax policy, redistribution of income may be only a by-product of what they consider an appropriate way of measuring the fairness of the tax system itself. Views about ability to pay, about comparative sacrifice, about comparative utility, and so on may be dismissed as resting on ethical rather than logical grounds. Blum and Kalven did a superb job of analyzing the difficulties with those modes of judging fairness in taxation, but in the end one does have to make a judgment about what is a fair tax system. One can make that judgment even on the basis of such vague concepts as ability to pay or sacrifice of utility without necessarily wanting to see income redistributed purely as an end in itself. That is why it appears to me that redistribution for some people can be described as a by-product of philosophical, ethical, or aesthetic judgments about fairness in taxation.

Norman Ture has, it seems, posited a rather different concept of redistribution by viewing it as an end in itself. But any tax system that requires one individual to pay more than another can be regarded as a redistributive tax system. Any tax other than a head tax is redistributive in the sense that it requires A to pay more than B. That's true even if it is proportional to income, to consumption, to real property, or to any other base that one might suggest for taxation. Consequently the mere identification of redistribution as a by-product of a tax system really advances the argument very little, unless Mr. Ture's point is that anything other than a per capita tax is redistributive and has some of the consequences of equalization. His primary concern is the trade-off between savings and consumption. But every tax, as he correctly points out, affects choices. Even a per capita tax would do that because one must somehow find a way of paying it, and that involves giving up something. For example, it will require those who must pay taxes to work, even though they might prefer not to work if they were not being taxed.

Musgrave: A reasoned and nonvisceral case *can* be made for an egalitarian approach in terms of consumption. If anything, the case should be made in terms of consumption rather than in terms of income. A good deal has been learned in that respect in recent years from the philosophers' discussions.

There is no necessary conflict between a gradual consumption egalitarianism and the freedom of choice between consumer products. Assignments in overall terms can be made, and then there can be freedom to trade. It is, of course, true that costs are involved in a redistributive operation—costs in terms of both efficiency and what in my paper I termed horizontal equity. I believe it is altogether wrong to think, as many economists do, that rigorous reasoning ends when the discussion progresses beyond the profit or utility maximization framework.

Graetz: I think Ture considers the market distribution somehow to be an ethical distribution. Since, given marginal productivity theory, the market distribution depends on demand, I have difficulty with understanding why it is an ethical result. I can see a justification for a market distribution as a concession to output and as a concession to certain liberties.

At least four liberties are directly involved. I think that Norman Ture and I agree that the liberty of the consumer to be able to buy "Muskrat Love" and the liberty of the Captain and Tennille to refuse to perform it—given what profit they would have left after taxes—are important liberties. Neither of us would stop the consumer from purchasing and neither of us would proscribe the Captain and Tennille from singing this song. But their liberty to keep what is paid by the consumer and the liberty of their heirs to continue to keep what is paid by the consumer strike me as somewhat different. They are not nearly so absolute. What is the justification for protecting the rights of their lineal descendants to enjoy an advantage produced because of consumer demand during the 1970s? I do not know why that liberty needs to be protected. And it strikes me that there is another important liberty at stake here—that is, the interest of lineal descendants of others to start off at roughly a position of equality of opportunity and equality of initial endowments. So it seems to me that we are back to balancing.

The liberty of lineal descendants to keep everything that people pay also strikes me as not absolute because it depends on things like their ancestors' natural endowments. The Captain and Tennille, if they were here discussing this with us, could have acceded to your request to sing "Muskrat Love," but I could not. I attribute that to a genetic difference in addition to a taste difference. The genetic difference that allows them to sing "Muskrat Love" and to earn all that money strikes me in Rawls's terms as morally arbitrary, and therefore is not a liberty interest that needs to be protected absolutely. There is a need, however, to make a concession to that interest because we do wish to protect the other liberty interests that I have mentioned, in particular the interest of the consumers to purchase this product.

Market distribution can only be justified as instrumentalist. The same thing can be said of certain redistributive policies. For example, if eliminating poverty is our only goal, that goal may be fulfilled through redistributional tax policy in part. And in that sense, a progressive tax can be defended on instrumentalist grounds, if not on absolute ethical grounds. The same thing is true in terms of compensating for externalities of failure to redistribute. For example, if crime exists because of a failure to redistribute, then some of these costs [of crime] may be alleviated by redistributing. A progressive tax may accomplish that goal, and an instrumentalist kind of defense of the progressive tax is again the result.

Orr: If the discussion is about the ethical basis of taxation and the fact of norms or goals, toward what are we being neutral? Are we being neutral toward the consumption patterns that would eventuate in a government-free society, or are we being neutral toward the utility benefits that are derived from total consumption of goods provided by government and goods provided by the market?

Ture: It is obvious in the context of my paper that the reference point is a taxless world and by inference a governmentless world. I would turn your question around. If a tax system is going to be constructed that distorts relative prices, there ought to be a fairly good reason for doing so. If in fact a tax is a tax, it has to raise the cost of resource utilization in the private sector overall relative to that of the public sector. But then the question arises of whether there is something that impels you to say, Let's raise some of the private sector costs disproportionately to others. I know of very little reference in the literature that argues that there is some social gain per se in raising the cost of future income relative to that of current consumption.

Galvin: Norman, let us assume that we have the tax system that you have described, that you have a rate structure that is acceptable to you, and that you have garnered the revenue that we need to run the government. Does it bother you aesthetically or ethically that those who sit in the power seats would, through the legislative process, take that money and use it for Medicaid and unemployment and welfare benefits and rent support? On the expenditure side, through conscious decision, will it bother you how we spend that money if we raise it by your method?

McNulty: It seems to me that we are having a cross-examination here, and what the cross-examination is showing is that you, Norman, don't care at all about equity, but you do care intensely about efficiency. To raise revenue, the anarchy and governmentless and taxless world that you would really like cannot exist. So (you say) let's have a tax that is the most efficient tax and has no price effect—such as a head tax. You have no logical or visceral inclination toward one kind of tax or another as to fairness. The end result is very often the division among lawyers and economists, with the lawyers worrying about fairness and the economists worrying about efficiency.

Graetz: I don't think it is correct to describe Norman's position as not embodying a view of fairness. I think Norman has a view of fairness. It seems very similar to a view that says people are entitled to whatever the market produces for them. It is consistent with everything I have heard him say about the similarities of his view to Nozick's, but without the procedural limitations.

McLure: I want to make explicit two things that I am sure a lot of people already realize. As long as Norman favors a flat-rate tax on consumption, he might as well simply favor a federal retail sales tax or a value added tax. The second thing is that we should recognize the implication of allowing a credit for saving under the expenditure base, instead of a deduction. The cash-flow income/expenditure tax that Bill Andrews has talked about involves not so much a deduction for saving as just leaving it out of the computation of income. It is quite a different thing, I think, to say that we are not going to leave saving out—which means we are still calculating income—but then to allow a credit for saving for certain people. Most of the simplification of the expenditure tax is lost once you try to substitute the credit for the deduction, if the availability of credit extends very far up the income scale.

Bittker: The standard of comparison of a "no-tax, no-government world" boggles my mind. If we move to a more realistic concept of a minimum government framework as a standard, what is needed is a specification of the legal institutions required to establish the standard of comparison that is implicit in Norman's theory. That is to say, presumably there would be a police force and a court system, plus a requirement that people comply with their promises. If so, do we hold people to promises, whether or not they were sufficiently educated to know what they were signing? A whole set of legal institutions require specification before one can envision this "minimum government world" that is then to be used as a standard to determine whether a particular tax system is neutral or not.

In addition, once we have established such a framework, it will almost certainly be enormously different from the world that we now have. Maybe it would support Mark Twain's view that no square inch of land is now possessed by its rightful owner. Maybe not. At any rate we would have to recognize such a disparity between that sort of minimum framework and what we have now. It is difficult to understand how someone like Nozick can speak of just entitlement when existing entitlements are based on incongruous legal institutions that the theorists themselves regard as erroneous, undesirable, or unfair. My point is that two problems remain. One is specifying the standard of comparison by which neutrality and departures from it are to be judged. Second, once the standard is specified, if it would result in a different world from the one we now live in, I don't see how the just-entitlement concept can be applied to the existing state of affairs.

Cooper: Do you mean that statement as a serious, political definition or as a philosophical idea? If the Sixteenth Amendment, for example, were mandatory rather than permissive, would you find that acceptable?

Buchanan: The attitude about the process is more important than the particulars of the institutions. It seems to me deadly dangerous for society, and especially for the preservation of a free society, if individuals start having the attitude that a big enough coalition can take rents away from another group that's in a minority coalition. I do object to every Congress, every year, jiggling with these rules of the game. It seems to me that these basic institutions must be conceived as ongoing rules that are changed in a different process and that we need a different way of thinking about changing the rules rather than transferring income back and forth within the rules.

Bailey: The ideal way to do things to get the best practical results would be, first, to imbed as securely as possible in the Constitution whatever rules we're going to have about redistribution so that year-to-year enactments on the part of Congress would be out of order on redistributive matters; second, to curtail the legislative activities of the court system, including the Supreme Court. Then the year-to-year process of policymaking in government would be a Lindahl-type process that would be solely concerned with efficiency. Such a rule would do what Dick Musgrave once proposed in a purely conceptual way; it would separate functionally the redistributive from the allocative activities of government.

Bias against Capital Formation

Feldstein: As to the question of the impact of taxes on savings behavior, three separate things get confused in much of the discussion: first, taxation alters choice; second, it distorts choice; and, third, the sign of that distortion or the direction of that change is unambiguous. Let me be more specific. Any tax may affect savings behavior, our current taxes distort savings choices, our current taxes reduce savings. Boris Bittker commented that even a lump-sum tax will affect choices, but it should be kept in mind that those are nondistortionary effects. That is different from a tax that puts a wedge between the price faced by households and the price faced by consumers. A wedge that distorts choices is the source of efficiency losses and of excess burden. That excess burden exists even if there is no change in actual behavior—that is, even if, in economics jargon, income and substitution effects cancel. Distinguishing those two propositions is needed: choices being altered and prices and incentives being distorted. The third point, the one I find most troubling, is the assumption that a tax that reduces the rate of return to savings necessarily reduces savings.

There is no particular reason why an increase in the cost of retirement consumption and the cost of future consumption should make one save less. Indeed, it may make one save more. I find not at all convincing the presumption that raising the cost of saving—what I like to think of as raising the cost of future consumption—inevitably makes a person save less. If the cost of future consumption is raised by taxing interest income, the effect will be that of making one want less future consumption. That's generally an accepted point, although it's at that point that income and substitution effects complicate the argument and make it ambiguous. A person who wants less future consumption will not necessarily reduce his savings; after all, to get a dollar's worth of consumption at retirement will become more expensive because of the taxation on interest income. A person will have to save more to get the same amount he would have received in the past. Consequently a person may increase his savings rather than reduce them. So the presumption in Ture's argument that our current system actually reduces savings certainly need not follow for households. That phrase "need not follow for households" is intended to leave open for later discussion the question of the timing of government deficits and surpluses as affected by the timing of tax receipts.

Finally let me say something about Norman's emphasis on capital accumulation. That problem interests me a great deal, but I am surprised that the entire emphasis on income distribution focuses on the capital side of the question. After all, the problem of redistributive taxation would exist even in the consumption-tax world favored by Norman, me, and several others. The same problems might be thought to arise in that tax world. I mention that in connection with, for example, the effect a relatively low income tax on the less skilled and a heavier tax on the more skilled, causing a change in the supply of labor by the more skilled.

I question Norman's continual combination of human capital and non-human capital, as if somehow the tax laws treated those two equivalently. Most of the cost of human capital formation is foregone income, and that's completely written off immediately under our current tax laws. If one decides not to go to work, but decides to spend another year in formal education, he is allowed to expense that in the sense that the income he doesn't receive is not taxed. The government operates a great many programs designed to increase human capital formation—public elementary and secondary schools, state universities which provide the majority of higher education in the United States, the special tax rules that favor private institutions of higher education.

Eisner: I would like to address the charges of bias in the current tax system against saving and against capital formation. These assertions seem far from correct. First, remember that not merely business acquisition of plant and equipment, but anything that contributes to the acquisition of a stock of knowledge, methods, techniques, physical plant, and equipment owned by the private sector, by government, and by nonprofit institutions, will contribute to future production. Second, the whole picture of saving looks like a crazy quilt. At the margin the bulk of us are paying very little, if any, tax on our saving, on our accumulation of capital, when it takes the form of contributions to pensions or increasing values of homes or other assets.

Major individual household saving takes place in the form of contributions to tax-exempt pension funds. What's more, the earnings received as these pensions are accumulated are usually tax-exempt. Considering further the various forms of capital accumulation and the ways in which people can accumulate income from earnings and from capital, we see an equipment tax credit of 10 or 11.5 percent, depending on whether one is dealing with corporations that contribute to employee stock ownership plans. With accelerated depreciation, the actual depreciation both for accounting and for tax purposes has most often exceeded economic depreciation. Then there is the lack of any taxation of unrealized capital gains and taxation at only half rate if gains are realized. Martin Bailey's calculations, now some years old, indicate an effective rate of taxation of capital gains at about 8 or 9 percent.

Untaxed income from housing also exists, which results in substantial incentives to the accumulation of capital in housing. Research and development expenditures are expensed, which means, in effect, that capital is accumulated without paying any taxes on the income that is involved in the production and accumulation of capital stocks of knowledge in the form of research and development.

The system remains biased against investment in human capital, particularly by those at the bottom of the economic ladder. But there is no taxation of the foregone income or opportunity costs of students, which are a large part of investment in human capital.

The argument about whether neutrality exists, must consider more than just the tax structure. Go back to the prime minister that Boris Bittker invented for us and ask him by way of comparison how much saving and capital accumulation he would have if there existed a society with no taxes, no government, and nobody to protect the capital that was accumulated. Then if a factory were constructed, somebody could just say, "Well, that's mine; you have no right to it and I'll take it away." In such a society I suggest that much less capital accumulation would take place.

The owners of capital get very large benefits from the huge national defense expenditures and from police expenditures that protect capital. If the United States were a poor country, it wouldn't have to worry about the Russians or anyone else taking everything away; they wouldn't want to. It may be argued that the major burden of the cost of defending capital should be paid by the owners of capital. Starting from that premise, then, one can build a very strong case that our system on the whole is biased in favor of too much capital accumulation. Recession slumps in business investment imply there is likely too much capital accumulation in the forms that Norm Ture is most fond of encouraging, like business purchases of machinery with equipment tax credits and accelerated depreciation for plant and equipment.

Bailey: I dislike words like bias or neutrality because they mean different things to different people. But there is an interesting problem in the difference between the private and social rate of return to saving and investment. Now, despite the catalog of things that Bob Eisner mentioned—accelerated depreciation and all that—there is no doubt that investment in tangible capital has a higher rate of return before tax than after tax. Therefore, there is the kind of discrepancy that affects efficient resource allocation.

Except for education and residential housing, most investment has a big discrepancy between the private and social rates of return. The most important single tax that creates this discrepancy is the corporate income tax. No matter how much depreciation has been accelerated, corporations still pay taxes, and their tax burden is a large part of the overall tax on the income from capital.

Eisner: I would dispute that. The corporate income tax is not a tax on capital. It is a tax on all the earnings of the corporation, and it is questionable that it changes the desirable ratio of factor proportions between labor and capital. In addition, the individual actually invests in the corporation with the notion of getting untaxed capital gains. So I question whether the after-tax return from tangible capital is less than it would be if no taxes were imposed and no capital gains exclusions were allowed.

Ture: Bob, on abstract conceptual grounds it seems to me you can very simply determine whether the base or configuration now in existence bears more heavily on income used to buy future income than on income used to buy current consumption.

Cooper: If there is any truth, as there obviously is, to Professor Eisner's statement that the bias against saving in the income tax is mitigated by various provisions, one also has to take account of the fact that consumption is not tax-free. There are sales taxes on consumption, and a portion of the corporate income tax is a tax on consumption.

Feldstein: A consumption tax is a neutral tax as far as savings go, as long as there is a tax on future consumption as well as on present consumption.

Eisner: The payroll tax is not neutral. The payroll tax is a tax on current labor as against the acquisition of capital used not only for current production but for future production.

Buchanan: Earl Thompson argues that what is precisely desired in this situation is the differential between the private and social rate of return on tangible capital and no differential on human capital. He calls the present tax structure ideal in the sense that the creation of tangible (or coveted) capital imposes a defense cost. In fact the defense budget can be attributed to that differential, and therefore the differential is desirable.

Eisner: I would add that there are other reasons for believing that the acquisition of human capital should be encouraged because of information costs and market imperfections that tend to discourage such acquisition.

Tolley: I question whether a consumption tax or an income tax is more favorable to human capital investment. It seems that an income tax is much more favorable to human capital formation than a consumption tax, because foregone earnings are in effect being expensed. If a consumption tax is in effect, however, and your earnings go down because of on-the-job training (which is about half of human capital formation, some people have estimated), this reduction of earnings cannot be expensed.

It is also questionable whether we should be concentrating so much on income and consumption in the marketplace. If we think of the concepts of full income and full consumption, the distinction between market activity and nonmarket activity is arbitrary. Choices are distorted when the tax system is based on what happens in the market. Tax bases that are quite different can be formulated. In some sense this whole discussion is unthinkable. We are not talking about what Congress is going to do tomorrow anyway. So, for instance, take a person's market wage rate, multiply it by 24 hours (because there are 24 hours in a day), and that's the person's full income. Figuring income this way assures neutrality toward market and nonmarket activity. Take full income and you will begin to get measures of investment and saving activity, both as they take place in the market and in the home. These would be guiding principles that would help eliminate some serious distortions, perhaps just as serious as the savings distortion in the household-market distortion.

Mieszkowski: The whole capital gains business, at least as it refers to corporations, has important bearing on the relative rates of taxation between corporate and noncorporate investment. Some arguments have suggested that when the typical stockholder is in a sufficiently high bracket, then most if not all of the earnings are retained. Then, indeed, the double taxation of corporate profit might be very small. But that certainly doesn't mean that there's an important difference between the average and the marginal rate of tax on corporations or that at the margin corporations and noncorporate enterprises are really paying the same for their cost of capital. That does not imply that there still isn't a substantial rate of tax, given the existence of the income tax and the fact that income taxes have to be paid on profits earned in unincorporated enterprises.

Eisner: There is a tax on corporations that I would eliminate, but my point is simply that the bulk of saving is undertaken by people who decide to buy stock or land. The effective marginal rate of taxation for most saving is simply not the rate of interest subject to tax that Norman Ture seems to have in mind.

Feldstein: I think Bob Eisner correctly emphasized what people hope for, expect, and anticipate rather than what they get after the fact. It's just not real estate investment trusts that have turned out to be less golden ex post than ex ante. Standard and Poor's index over the last 20 years has risen just about as much as the consumer price index, so that somebody who, in anticipation of continuing growth of share prices, bought stock ten years ago and sold it today, would discover that his real value hadn't gone up at all. Nominally his value had doubled, and his tax on that essentially wiped out the small after-tax dividend income that he accumulated over the period, leading to a zero real net rate of return on corporate investment. Bob Eisner said that the corporate income tax falls not just on the capital of the corporations, but on the entire corporate income. Unless the concern is only the general shifting point, then I think I disagree. Although it is indeed a tax on corporate income, it's a tax on corporate income defined after allowing for the subtraction of wage payments and of debt interest payments. So it is a tax only on equity capital income.

We go around in circles a little bit on this point because we cannot seem to decide in these discussions whether we are talking about supply or demand of capital. If national savings are fixed and the labor supply is fixed, then changes in taxation are going to affect the after-tax return, and they are going to affect the allocation of that capital stock and that labor stock between different uses. But obviously, by definition, they are not going to affect the total amounts. So the tax on capital in the corporate sector does not discourage the investment of capital. It just changes the rate of return under the assumption of fixed savings.

When you talked before about payroll taxes discouraging the hiring of labor, you implied that the supply of labor is variable. If the supply of labor were perfectly fixed, then there would be no change in the amount of labor that corporations would want to use because of the payroll tax. It would imply, instead, a fall in the net wage that labor is paid.

Tolley: Speaking more directly to the incidence of the corporate income tax, I thought there was quite good agreement that the corporate income tax is a tax on all capital. It's not a tax on labor, and it's not just a tax on corporate capital; it is spread to all capital. I also thought that everybody agreed that it does discriminate against equity financing relative to debt financing. It has been my impression that most of the fraternity agrees on that.

Eisner: I would have the combined government and tax effects on saving be nonneutral only to the extent that I felt that imperfections in markets or externalities were taking us away from the revealed preferences—the choices of individuals. That means that I would probably, as a first approximation, be happy with a combination of a consumption tax and a tax on capital—a tax on capital to reflect the benefits received by society from government for protection of the capital.

Capital Formation and Growth

Meiselman: What is the relationship between capital formation and economic growth, real wages, and employment? What might be learned from the experience of other countries, Britain on the one hand or Japan, where there has been a very high rate of saving and investment, on the other hand?

Eisner: Although I trust it will impart nothing new to most of the economists and perhaps to many of the lawyers, I would like to correct the idea that capital formation is holy because it creates economic growth. We should be free to choose the portion of our income that we want to devote to current consumption and the portion to the future. There is no reason to encourage capital formation, meaning future consumption, at the expense of current consumption. More capital formation in the neoclassical analysis does not mean a higher rate of growth. In equilibrium it will mean a higher path of output to the extent that there is a positive net marginal product to capital. That positive net marginal product does not necessarily exist for each type of capital at every point in time.

This means that measures to encourage capital formation will bring a higher net output only to the extent that such measures create additional capital and only to the extent that such additional capital adds net output. I give laymen the example of a businessman who considers the acquisition of a piece of equipment that costs $100 and would contribute $95 to output over its lifetime. Giving a 10 percent credit to buy the machine is not going to add to the social output. So capital formation does not necessarily add to net output, let alone to growth.

As far as wages go, a production function can be assumed where more capital means a higher return to labor, but these things depend on the shape of the production function. As to capital formation and employment, moves to try to induce the acquisition of capital in an economy with no guarantee of full employment do have a substitution effect that operates against the employment of labor. It is true that the production of capital goods requires additional workers, but then the production of any additional output requires additional workers. There is nothing uniquely desirable about encouraging capital formation for employment. Because of the substitution effect, that would seem to be a less efficient way of bringing about employment than other stimuli to output and demand.

Bailey: I would like to offer a rejoinder to part of what Bob Eisner just said. If the fraction of income saved by any means is increased, and if it has a positive marginal product, then the rate of growth is increased. The rate of growth times the capital stock can be defined as equal to the amount of annual investment. If the amount of investment is increased, the rate of growth of capital stock is also being increased. When the saving fraction goes up, the rate of growth of the capital stock rises and the rate of growth of income rises, assuming that the rate of return is positive.

Musgrave: But with constant technique and other factors, the growth rate would be limited by the rate of growth of population because the return to capital will decline if accumulation moves ahead of population.

Bailey: I accept your correction in the sense that the rate of growth of the capital stock to which the economy converges is unaffected by the saving rate, if a classical growth equation with no technical change is applied. The economy converges to a higher path, and for every finite period the rate of growth is higher because of the convergence to a higher path.

Let me comment also on the question of optimality. Two reasons have sometimes been stated to explain why a rate of saving and investment that satisfies the ordinary prima facie optimality condition might not be right. By the prima facie condition I mean that the rate of return received by the saver is equal to the social product of the capital (the pretax rate of return on capital). However, there are two interesting possible exceptions. One occurs when capital yields satisfaction other than its flow of income, so that the amount of wealth would enter into people's utility functions. The other reason, which Bob mentioned, is that capital might have to be defended. The additional capital would add to the total defense cost of the country.

As to taxing the consumption value of capital, the point is unquestionably right in principle. If the Kaiser's lieutenant likes the opera, he should be charged a little extra income tax if his duties require that he go to the opera; and if he doesn't like the opera, he should be given a deduction, I suppose. If people accept lower earnings than they could get in some less pleasant place or environment, they should be charged income tax on the consumption value of the pleasant environment. Such a consideration is nice to state in principle, but it seems to me that it has no practical application. Moreover, so far as I can see, there is virtually no way to test the hypothesis that wealth enters the utility function, especially that it does so at the margin. Many people may get direct utility from owning wealth, yet it might have no marginal effect whatsoever.

On the defense issue I think the point could be true, but I doubt that it's of any great size. Given the relatively small part of gross national product that is devoted to defense, the appropriate tax on capital to finance its share of its own defense would, I think, be rather small.

Meiselman: Isn't the point about the importance of capital formation related to the discussion of the bias in the tax system against saving and investment? I don't think anybody would argue that more capital formation is better without any limit, because at the limit we wouldn't have any consumption.

Feldstein: Let me start by agreeing with much of what Bob Eisner said about being free to choose (indeed, essentially the market should be free to choose) the amount of capital accumulation undistorted by government policies. But, as David Meiselman just said, what we've been talking about is that the current tax laws do distort the rate of savings, and so it's not appropriate to say that any outcome of the market process is just as good as any other outcome. Outcomes of the market process distorted by particular taxes are less good than undistorted outcomes. So what we're really talking about is trying to get back to the outcome that would occur in the absence of the current tax distortions.

The suggestion that particular investments may have negative productivity is interesting, but I think in the aggregate no one would argue that the marginal product of capital is zero now. Even low estimates of the marginal product of private capital would run around 10 percent or higher.

Finally the statement that the effect of capital formation on real wage rates depends on what is assumed about the production function is true quantitatively, but I doubt that it's true in any meaningful sense. That is, any economic specification of the production relation that I've ever seen an economist use assumes that an increase in the capital stock increases the marginal product of labor and therefore real wages. Whether one assumes a Cobb-Douglas technology or some other technology influences the *amount* of the increase in wages associated with the given increase in the capital stock, but not *whether* such an increase in wages would occur.

Mieszkowski: The first point that is of ultimate interest is not capital formation for its own sake, but its impact on consumption or potential consumption. Bob Hall shows that a switch from an income tax to an expenditure tax would have a dramatic effect on the amount of capital formation and on the rates of return, but the effects on consumption are much less dramatic. If an increased output is desired, given the constant population growth, it is necessary to be more capital intensive. It is imperative to save more and consume less. It's as simple as that.

Tolley: Denison started out to answer that question about the sources of growth, and he did get some very rigorous quantitative measures. Capital formation accounted for about a quarter of the growth. Physical capital formation is quite important, but other forms of capital formation (human capital, research and development) are more important. That contains a lesson for people who are interested in taxes. Tax analysts have not yet given as much attention as they should to these nonphysical forms of investment.

Ture: For some relatively small increase in the capital stock (and that's all you can be talking about), a move toward the most nearly neutral saving-consumption tax system implies only a relatively small increase in the real wage rate above the trend value of growth in real wages. But there will be an increase in wage rates and in employment relative to the values that would otherwise be realized.

The more significant question is why we should have a tax structure that prevents the market results of saving and consumption choices from being relatively undistorted. Let us have a tax structure that avoids that kind of distortion, and let us happily accept the resulting rate of increase in real wage rates, in employment, and in real output.

Tolley: I really do believe that we should be trying to get more quantitative on these issues. How important is capital formation to employment? I believe—and I think many agree with me—in the long run it has nothing to do with it. If we're thinking about the tax structure as a long-run question, then we should not be concerned with the short-run employment objective.

Feldstein: If the coefficients of the Cobb-Douglas production function are, say, one-third for capital and two-thirds for labor (standard numbers implied by the factor shares), then a 1 percent increase in the capital stock leads to a 0.33 percent increase in total output. Since the shares of capital and labor remain unchanged at any level of output, regardless of the capital-labor mix, if output goes up by 0.33 percent, so, too, does labor income. Now, raising the wage rate by 0.33 percent sounds very small, but that's misleading because, as Norman Ture said, one shouldn't think of it in those terms. Nor should one think about it in growth terms because, as Bob Eisner said and Martin Bailey agreed, in the long run there is no growth effect. The real question to ask is, What is the rate of return to an additional dollar's worth of capital formation? The answer, I think, is about 12 percent. Consequently we must also ask whether at the margin that's a high enough rate of return to indicate the necessity of more savings. One way to answer that question is to ask whether individuals are choosing to save and thus facing that price of 12 percent as the rate of return. Or is that the rate of return available to society but not to individuals? The effective rate of tax that individuals pay on capital income on average is about 43 percent, according to the calculations implicit in Norman Ture's calculations. The 43 percent tax means that a 12 percent gross rate of return on additional investment corresponds to a net rate of return somewhere on the order of 7 percent. So the distortion in savings comes because society can earn 12 percent at the margin and individuals on average get only 7 percent.

Meiselman: As long as you're our resident expert on the numbers of the Cobb-Douglas production function, would you care to comment on some of the potential employment implications, not in a world where everything is flexible and the outlook is over a very long period of time, but in the current labor market?

Feldstein: I think being the resident expert on the Cobb-Douglas production function doesn't help in that situation. I believe that there is no long-run employment effect.

Meiselman: What about before you get to the long run?

Feldstein: I can see it going both ways. I don't know how to determine what the net effect is.

Musgrave: The labor supply function exists in terms of real wages, and the demand for labor depends on its productivity and the real wage rate. Capital formation and growth would affect the real wage rate because it would affect labor productivity. This would result in an increase in hours worked at zero involuntary unemployment, but it would not affect involuntary unemployment in the long run.

Eisner: It is clear that with most production functions, in the aggregate, a higher amount of capital will mean higher income to the other factors or production. In a two-factor system, that is precisely what is going to happen. Nobody here can possibly believe that all additions to capital raise all real wages. Unions know that certain kinds of capital accumulation put certain kinds of people out of work and reduce their real wages. If you want to raise real wages, the obvious thing to do is to increase the capital that goes directly to producing labor income—and that is essentially human capital.

Sunley: I thought there was general agreement that capital formation in the long run has no effect on employment. Some people, however, have made a career of estimating for tax-writing committees of Congress what will happen to employment if tax shelters or some other tax subsidies are cut back. Tax policy is usually made on the basis of the short-term effects. If there were two proposals on the table, both costing about $2 billion, which would you vote for if the purpose of the short-term program were to stimulate employment? The first proposal is a tax credit equal to 5 percent of the employer's share of FICA, and the second proposal is a 2 percent increase in the investment tax credit.

Feldstein: I would have to know more about the facts to answer the question. If the assumption (perhaps incorrect) that we have in mind in the short run is that there is a fixed amount of labor, then the case for the investment tax credit would be that investment goods-producing industries are running at very low capacity. If these industries have a lot of unemployment, then a selective boost to demand in those industries will be better for employment than a general reduction in the cost of labor in all industries. So my conclusion would depend on what I saw in the structure of current unemployment.

Unidentified: To some extent, Marty, that's actually what Arthur Okun calls "the-penicillin-in-the-throat" theory. It's true that in a downturn unemployment goes up in the capital goods industries areas more than any other place. It's a much more cyclical industry, and so advocates of increases in the investment credit come in and plot machine tool orders over the cycle. They find out that the machine tool orders are in bad shape because of high unemployment, and they conclude that something is needed to stimulate investment, because that's where high unemployment exists. Now, when I have a sore throat, I'm perfectly willing to take my penicillin in my fanny if that's where it's needed, hoping that it will work its way through the system. But economists never seem to figure that out, because they feel that if the throat is sore, the penicillin should go down the throat.

Feldstein: I'm not sure about the technology here. The analogy, while amusing, may be wrong. If you are an elected politician whose term in office will be over after 12 months and you're going to be evaluated by history or the voters in terms of what you do in those 12 months, then you may be a good politician, but you're a bad public servant. If you have to choose, given that horizon, it may be better to take the penicillin in the throat in this case. Penicillin may work more effectively in your fanny than in your throat, but directing your tax policies at the industry that matters—getting a price effect as well as a demand effect—may be a more effective short-run stimulus. It may make very bad long-run policy. It may be that distortions are much worse, they may be better. But you asked me to focus on the very short run, and I think from a very short-run point of view, it probably does pay to target things where the problem is greatest.

Musgrave: In the very short run, the reduction in the wage rate that would result from this proposal would be infinitesimal. It would be less than 0.33 percent, whereas the investment credit increase would have a significant effect on the rate of return. From the point of view of announcement effect, the payroll deduction approach would be much less significant. For that approach to be significant, it would need to be much more substantial for a much more limited type of labor. But, as the proposal now stands, I think it makes no sense at all.

Hellawell: I just want to add a footnote to Professor Musgrave's comment that the low-income economy can less afford the efficiency costs that may be involved in redistribution and can less afford to forego savings. Even today, in a large part of the world, redistribution would put everyone in the society at a bare subsistence level. Until quite recently that was the case in almost all the world. Where egalitarianism would lead to a bare subsistence level for everyone in a society, that society could not even record its history or develop its culture without a nonegalitarian distribution of income. In those cases, which are not uncommon even today, there may be an aesthetic or even a moral reason for unequal distribution of income.

Redistribution

Bailey: In principle the entire government could be supported by benefit taxation so that it would have no redistributive impact whatsoever, in the sense that paying taxes would be like paying for value received, as in the private sector. If the existence of government had no appreciable effect on the standard of living that we could enjoy relative to a state of anarchy, the government would have to operate this way. One cannot redistribute if people opt out and avoid the redistributive effect without cost to themselves. But given that large rents are enjoyed from the existence of organized society, there is some scope for redistributive activity relative to a Lindahl equilibrium or pure benefit principle for support of the government. Experience shows that almost everyone is tempted to exploit the opportunities to tax some people's rents in order to supplement other people's rents.

Regarding the ethical principles that discussions in this conference have tried to bring to bear on this issue, such as Rawls's maximin, my own personal predilection is to accept Mark Twain's judgment about these matters—people have been up to skullduggery for so long now that not a square inch of the surface of the earth still belongs to its rightful owners. And that being the case, I can't get terribly excited about the urge that most people seem to have to indulge in redistribution. If a political movement can be organized and can assert itself successfully, then that creates some new property rights or a new distribution of the present property rights. However, it appears that the problem will come full circle. I get upset when people are willing to sacrifice some income themselves in order to reduce someone else's income. At that point it becomes immoral.

Buchanan: I share with Norm Ture a distaste for the redistribution aspects that a lot of the modern discussion carries. People propose to use the instrument of the state to take away from some people and give to others and to manipulate the instrument of the state. I think we should consider how to build into the structure of institutions adjustments on the sharing mechanisms of the product of the society. That is quite different from saying year by year, period by period, that one coalition takes it away from another. It seems to me that the basic structural adjustment was what John Rawls was getting at. It seems that some attempts to modify or change the sharing structure must be built into the structure of basic institutions, and that is quite different from using the state for one coalition to exploit another coalition. The market allocation has no ethical properties as such.

Ture: Why do we have to build in an institutional arrangement for altering the sharing of the product of society anymore than we have to leave it the way it is?

I don't know how to make the link from the premise of the equal, moral worth to the notion that it calls for anything whatever either by way of ad hoc or permanent and enduring arrangement for changing the distribution of outcomes.

Buchanan: What emerges by happenstance has no more ethical support than any other alternative.

Ture: But it has no less ethical support, either. It does happen, subject to some observable rules with respect to the determination and exercise of property rights. But you are saying that those outcomes are in some sense unacceptable and should be altered. Why not accept what befalls us?

Orr: I would like to interject the explicit idea that ethics is a normative discipline concerned with relations among human beings. The egalitarian impulses that we have been discussing can be broken down to two types. One, we can talk about provident impulses, feelings for less fortunate people, feelings for individuals clearly worse off than ourselves. There is always an option to exercise such impulses through spontaneous, unilateral voluntary actions. I find the objection that this kind of mechanism is ineffective on historical grounds less than wholly persuasive in the context of society today. I think the objection that there is a big opportunity for free riders in such a form of organization is just wrong. A dollar given by an individual to somebody else to alleviate that individual's misery has accomplished its purpose.

The second impulse that we can talk about in addition to the provident impulse is the insurance impulse. Individuals want to protect themselves against disasters, to protect themselves against misfortunes by creating mechanisms they can rely on in the event something does befall them. I would like to stress that the state is not necessary as an agent in the exercise of either impulse. The state may be an efficient agent in providing the kinds of service we are talking about, but it is by no means a necessary agent. Now, keeping in mind that ethics does bear on relations among individuals, the objectionable feature of redistribution, for provident or insurance purposes, is the coercive element. One is forced to participate in the process of redistribution through the police power of the state. I think this tends to blunt or deflect some of the virtue of the ends that are perhaps being served. If I may be a little bit hyperbolic, I would like to characterize our current system as one in which individuals are coerced into membership in a morally neuter and administratively corrupt surrogate for the Church of Jesus Christ of Latter Day Saints. That is, the welfare system in America today has objectives that in large measure overlap the Mormon Church or its membership. But on the one hand we are dealing with a voluntary and fairly effective mechanism; on the other hand we are dealing with a mechanism that breeds considerable resentment and is administered in a very inefficient way.

Musgrave: As I have argued in the paper, I think it incorrect to look, as Ture does, at the cost of redistribution in terms of the reduction (or lesser growth) in earnings that may occur in the future. This view ignores the fact that reducing saving increases consumption at an earlier period. The real social cost is the efficiency loss or the deadweight loss that results from that interference, and not the fact that the level of future income is lower. Nevertheless, reduced growth is to the disadvantage of future generations. But, given our technological progress and given the tendency for population growth to decline, I am not too worried about the increase in the level of per capita income that people will have in the future.

Meiselman: I would like to direct the discussion to the feasibility of income redistribution and particularly to the question of the existence or the nature of any trade-off between the level of income or the rate of growth of income and the attempts at redistribution.

McNulty: Alice Rivlin, in a speech published by the Brookings Institution titled "Income Distribution: Can Economics Help?", points out that statistics indicate that there has been little or no apparent income redistribution or wealth redistribution in the last 30 or 40 years. Perhaps that can be accounted for by flaws in the statistics and countervailing forces. This was during a period when we have had a nominally progressive income tax, a regressive social security tax, a progressive estate and gift tax, and some other tax and transfer packages. She points out that the multiplication of family units, as young people tend to emancipate themselves earlier and old folks tend to maintain their own homes rather than living with the middle or younger generation, may distort the statistics, which are computed by households rather than by individual family members. But still, she is puzzled by the apparently small amount of redistribution. Perhaps this may be explained by the Pechman curve showing that the income tax is much less progressive than nominal rates would suggest.

I wonder about the possibility of giving up the notion of using an income tax with graduated rates for purposes of redistribution. Perhaps we should adopt some of Ture's notions, such as resorting to a different tax package, particularly one in which the marginal rates of the income tax would be proportional, uniform, and applied to a comprehensively defined base. I understand that there is some economic literature showing that the optimal tax might well be a proportional tax. When I have asked economists why that is true, one answer I get is that the proportional tax rate would retain incentives for highly skilled, high-income people to work, invest, and take risks and that the net effect might therefore be as Ture suggests—greater wealth for all and greater ability to achieve a redistribution of wealth.

Ironically, we might get a better redistribution than we have been able to achieve in the past years, if that is what we want, by an income tax that would appear to be less likely to redistribute wealth. Or it might be better not to try to redistribute through the tax system at all—to handle such redistribution as we want through the positive transfer system. Perhaps government should get out of the business of trying to redistribute, except to guarantee the well-being of the blind, the disabled, and the mentally retarded—those who are utterly unable to provide for themselves—and perhaps should rely as much as possible on private charity. Maybe because we have been so unsuccessful so far, we will have to surrender the notion that we can obtain redistribution through the tax system, whether or not one thinks it desirable as an ultimate policy goal.

Beyond that, my own work shows we can have even more redistribution and actual progressivity in the federal tax system if we have a flat-rate (proportional and uniform) individual income tax rate schedule applicable to all income (including capital gains), repeal the estate and gift tax, tax gifts and bequests as income, fully integrate the corporate and personal income taxes, repeal the social security payroll tax, and substitute an income-tax-financed universal social security payment or demogrant for welfare programs and present social security payments. The most important point is to see that a proportional rate at the margin can be combined with exemptions or demogrants or an exemption and a negative income tax to produce a very progressive income tax. The rate change would reduce transaction costs and undesirable incentives in a magnitude to delight any of us.

Orr: I would like to comment on the use of the statistical abstract quintile income shares as an indication of income distribution through time in the United States. There are myriad reasons why intertemporal comparisons using this indicator are meaningless, not the least of which is the reason Marty Feldstein pointed out in his work on social security. If we impute entitlements to social security's future beneficiaries, then in fact wealth is much more uniformly distributed than the quintile share indicator shows. The demographic changes through time have been absolutely devastating to the problem of intertemporal comparisons. One particular consideration is that the age distribution has changed markedly. In the 1930s something like 6 percent of the population was over 65. In the 1970s something like 13 percent is. Furthermore, because of social security, those over 65 are living in their own households and are counted separately, whereas before they were counted as part of the households of their offspring who were caring for them. So getting all excited over the glacial drift in the quintile shares is an exercise in futility. We can point to these quintile shares as a basis for our intuitive or visceral feelings that things are too unequally distributed. But to conclude that past policies have not done a job of redistributing income, for better or for worse, is absolute nonsense.

McNulty: One possibility is that the American people don't want redistribution, not even the people we think would most obviously want it. There have been proposals in California, for example, for the confiscation—or at least very high taxation—of estates above some level, say half a million. However, it turns out that people working at moderate salaries say they don't want such a law. One explanation is that some of them think they're going to be in that monetary position or that their kids may be, and they don't want to be hurt by that kind of legislation. Another possibility is that the entertainment value of the affluent is worth something to the poor. Also it may be that they do recognize the incentive and reward values that the radical proposals would tax away.

Orr: Or maybe they care about property rights.

McNulty: It may be that. It seems an important paradox that over the last umpteen years, with the one-person one-vote rule more or less in effect, we haven't had radical redistribution of wealth—because voting power is there to do it. And it seems to me regard for property rights and maybe some intuitive feeling about incentives may have kept the mass of the voters from taking from the minority or the very well-to-do and redistributing among the masses or the very poor. Why hasn't that happened? Is that possibly an indication that the people at large don't want massive redistribution? Or does it merely reflect on the political process?

Graetz: I would like to note that both a Democratic Congress and Ralph Nader's tax reform group, each of which you might expect to be predisposed to redistribution, endorsed just within the last few months a one-third reduction in revenues from the estate and gift taxes. The studies of Jim Buchanan and George Stigler and others using a self-interested view of government suggest that one is likely to find redistribution that serves the middle class.

If you look at the expenditure side, you're going to find a lot of redistribution from both upper and bottom brackets to the middle class.

Ture: With regard to the redistributive effects of the tax and expenditure sides of the budget, as I tried to illustrate in a little model I put together in the paper, what you essentially get from the tax system itself is very little by way of income redistribution. The redistributive effects hinge on what the expenditure system looks like. In the postwar era the developments on the expenditure side strike me as representing a very strong movement toward an increasingly redistributive expenditure system.

Having made that observation, I want to accede to what Dan Orr was suggesting. The kind of information that we typically use after transfer distributions, for measuring both pretax and after-tax, is miserable stuff. It isn't miserable merely as a conceptual problem, as Dan suggested, but those who have engaged in the process tell me it's also miserable in terms of the way in which the data are collected. What you have is information about the distribution of income to which nobody should pay attention.

McNulty: Let's assume for the moment that we're stuck with an income tax or that the transition to a consumption tax would take quite a while. It is arguable that, for reasons you gave in your paper, Norman, a proportional-rate income tax would be likely to provide a bigger pie, and therefore greater welfare for all. And with exemptions or an exempt demogrant, or a negative tax feature, it might provide greater redistribution than the tax we have had, which is nominally progressive, though eroded at the base and flawed in lots of other respects.

Ture: What I have been trying to point out is that the difference between going one tax route versus another tax route is primarily a matter of how large the pie will be. It is not primarily a matter of how that pie will be divided.

McNulty: And mightn't that pie be larger if the marginal tax rate were proportional and uniform, rather than multivaried and graduated?

Ture: Sure, because one way of looking at it is that the progressive rate structure simply represents a penalty tax on increasing one's productivity. It certainly strikes one as counterproductive socially.

Musgrave: I very much agree with what Norman said about looking at the redistributive implications of the fiscal system as a whole, including transfers and not only the tax side. Changes that have been distributionally significant in recent decades have been altogether on the transfer side. Looked at in that way, the share of the lowest decile in the distribution has been increased very greatly through the redistributive process. The share that low-income people have in terms of disposable income is many times as large as their share in earnings only. In that sense redistribution has been very significant.

Now, on the other hand, if you look at what happened to distribution as a whole (the Gini coefficient), you will see that very little has happened. So I think the objective is very important. Using a mostly flat-rate tax and obtaining one's redistributive effects on the transfer side may get one much further if the focus is on what should go to the very low end of the scale. But if the concern is with not having too large a share in the upper quarter, then, of course, a flat rate won't do. So it's important to decide whether the major focus is to raise the bottom share (reduce the Gini coefficient), or whether it is to look mostly at the top quarter.

Orr: The overall pattern of taxation and expenditure by government has an important bearing on the welfare of people at every level of the income distribution spectrum. In recent years one form of taxation has been inflation. I think that low-income people tend to be hurt by inflation to a greater extent than higher income people. This is an example of the way that expenditure policies, fiscal policies, and monetary policies interact with taxation policies to distribute well-being. If we take seriously the idea that income is too unequally distributed or consumption is too unequally distributed, then we can in fact suggest ways to remedy the situation. But those suggestions must necessarily include proposals for expenditure patterns as well as revenue collection patterns. And any proposals that serve that particular objective will probably have important effects on other objectives, such as the stabilization of employment in the short run. A lot of work will be necessary in order to figure out what kinds of trade-offs are involved and how in fact they should be reckoned.

Ture: Surely it is not obvious a priori that the nonworking poor are more seriously injured by inflation than all the rest of us. They do not bear the explicit income tax incremental burden, because by assumption they don't pay income taxes. The empirical question would be whether or not the welfare clientele have received increases per client in their various transfer payments at a rate equal or less than or greater than the rate of inflation, since almost all the programs are now indexed.

Orr: The answer that welfare benefits are indexed is a curiously out-of-character response from Norman. I think that one can say, "Okay, this poor drudge has been put out of work by inflation and a progressive income tax. Now he is on welfare; we can forget about him." That, in my view, is not a good response to my contention that he is differentially disadvantaged by inflation as a form of taxation. I think if we really want to take John Rawls more seriously than someone to whom we genuflect in discussions like this, we have to look very carefully at his emphasis on the means of self-respect as one of the primary social goods in any society.

McNulty: I think that Norman Ture made a useful distinction between the distribution of the tax burden and the posttax distribution of income. He went on to say that the tax law contains a bias against savings and therefore against income for investment. But the federal income tax, in some respects at least, contains a substantial bias against income from labor. There are some offsetting provisions, I agree. But if there is a bias against earned income, that means in effect a bias in favor of income from investment or savings. It seems to me now that the difficulty arises when one tries to keep the separation. That is what I would have liked Norman to emphasize. The question of the fairness of the tax burden has something to do with the source of the income or the wealth, whether it was earned or unearned, whether it was from a government transfer, or from a very clever and socially useful invention, or from theft. Yet, having decided that the tax burden ought to be distributed a certain way because of the source of the income or wealth, an egalitarian might then want to look at the after-tax distribution and say that although the tax distribution had been fair according to some moral or other standard, the after-tax result was somehow intolerable and that more (or less) equalization should result. So I urge that we attempt to keep a genuine separation between the notions of the fairness of the distribution of the tax burden and the fairness of the posttax distribution of wealth and economic power.

Goetz: As we begin to talk about government and politics in the discussion, I would like to ask why anyone might be interested in the before-tax distribution of income, as Norman Ture seemed to be in his original paper, for instance. I think the reason we might care about the before-tax distribution of income has a lot to do with philosophy or ethics or perhaps one's view of governmental process. It goes to the question of whether you want to change the world, once and for all, with your redistributive process, or whether you want to keep adjusting it within every period for all times. If, for instance, we posit a world in which there are no costs of transferring income after tax, that's the functional equivalent of lump-sum taxes which might be different for different people. But there are no excess burden losses. Then we can go on constantly just worrying about the after-tax income. But even if we eliminate the possibility of the economic costs, someone like Jim Buchanan would worry because the mechanism that is supposed to be redistributive necessarily becomes infected with a struggle between different groups over goals that are redistributive, not in the original Rawlsian sense, but in year-to-year strategy. The trade-off can be seen between trying to change the world—that is, to change the distribution of capital and to change the distribution of before-tax income—and adopting a procedure in which some kind of costs are accepted in every period to make adjustments in those periods. That's an important distinction to make.

Andrews: Like Ture, I am not very attracted by a picture of total equality and homogeneity of consumption or anything else. On the other hand, I'm quite disturbed about extreme inequality of consumption. That's exactly where the concern ought to be. It's people who are starving at one end of the scale, and it's people who are pursuing one kind or another of excessive, conspicuous consumption at the other end of the scale. If the inequality that we want to relieve is an inequality of consumption, then the direct way to deal with that is through a progressive tax on consumption with exemptions at the bottom. I think you have to have transfer payments at the bottom to deal with one side of the problem, but I also think you have to have the tax going up to quite healthy rates at the top, too, if you think that is part of the problem. As for wealth, income, and different modes of participation in the productive sector of society, we ought to welcome all the disparity that arises. That is, if there are some people who like to hold tenure at a university and enjoy the opportunities tenure gives them, and there are other people who enjoy power through participation in the political process, and other people enjoy military office, and other people enjoy manipulating golf balls, and other people enjoy manipulating large aggregates of capital through making production decisions and anything that goes with it, I think we ought to encourage that diversity and not try directly to equalize that at all, but to address ourselves to the consumption end. So I end up thinking that redistribution may be much more feasible economically if it's pursued directly in terms of reducing inequality in living standards.

Consumption Tax versus Income Tax

Andrews: Why doesn't one say directly that consumption is the thing that ought to be redistributed, altered, not equalized, but made less unequal, rather than income or wealth?

Wolfman: I agree with the objective of reducing the extremes of inequality in consumption. Yet, I am tentative, worried, and hesitant about the notion that the goal shared by some of us can be achieved by moving to what we call a consumption tax. It is perfectly possible for people to see that what we are calling a consumption tax is just an income tax, such as we now have, but with a new deduction—a deduction for all savings. And once we have an income tax with a deduction for all savings, we may still have a senator who will get up on the floor of Congress and say, "You know, it's a very good idea to help cities and we want particular savings encouraged, savings that will help cities." So he will come up with an ingenious new deduction, a deduction for consumption to the extent that it's provided by income from municipal bonds. That's just one step toward erosion; and there is no reason to assume that if a deduction for savings is allowed, a very expanded, comprehensive income base will automatically result.

Brennan: It's one thing to argue that a consumption base is better than an income base. Suppose we agreed on that. It's nevertheless not the same to argue that we want to move from an income base to a consumption base. There are two reasons why we might take that position. One is the question of transitional equity. The changeover itself would involve income redistributions that have to be taken into account. It may be that these transitional problems can be handled, but I think they need much more attention than we have given them. The second point is one to which we have alluded in various ways during the discussion. That is, in some ways the tax system is quasi-constitutional. It sets the context within which decisions about public expenditure levels, private contracts of various sorts, and so on are made. There are costs in changing the Constitution too often. It seems to me that in some cases it may not be so much the nature of the law that's important, but the fact of the law—the fact that the law exists. There are lots of examples of this, simple laws like the one that tells us which side of the road to drive on. If one imagines that law to be determined, for example, by the toss of a coin each morning, then the implications are disastrous. There is an element of that in tax reform. A lot might be gained if we all agree to a moratorium on tax reform for a while.

Feldstein: Of course we economists are all rank amateurs at trying to guess the political effects of consumption base versus income base. But some of the arguments, both good and bad, against progressive taxes and broader bases that are stopping us from getting the kind of income tax that we would like wouldn't apply to a consumption tax. One thing that would make me accept higher effective tax rates for a consumption tax is that all the arguments suggest that a consumption tax has a smaller efficiency loss and a smaller excess burden per unit of revenue. That's a good argument, but it's not likely to be a persuasive political argument. Political arguments, it seems to me, are of the sort, "If high-income people are taxed at high rates, saving will be discouraged. Saving is important for capital accumulation; therefore, we must not tax them at high rates." That kind of an argument, which appears more desirable politically, would be completely eliminated by going to a consumption base because the claim could no longer be made that a high tax on high-income individuals would discourage saving; their savings are completely untaxed. Indeed, if anything, the higher the tax, the more incentive they would have to save.

Brennan: It is possible that the optimal tax requires future consumption to be taxed more heavily than current consumption by virtue of the complementary relationship between future consumption and leisure? But suppose we genuinely don't know what the relevant substitution effects are. Then the relevant question is, Is a tax system that introduces an arbitrary distortion preferred to one that does not? And the answer to that question seems to me to run as follows. Although it's true that introducing an arbitrary distortion is just as likely to move us in the right as the wrong direction, the costs of moving in the wrong direction systematically exceed the gains from moving in the right direction. If we think of the relevant welfare triangle, we know that the gains from moving in the right direction are given by the welfare triangle. The costs for moving in the wrong direction are given by the area under the demand curve over the range that systematically exceeds that welfare triangle. Therefore, it seems that there is an a priori efficiency case for not introducing an arbitrary distortion, such as the one between present and future consumption, when we have no information about the relevant substitution effects between present and future consumption.

Musgrave: The statement that there is no a priori case grants that the income tax interferes with (1) substitutability of present and future consumption, (2) present consumption and leisure, and (3) future consumption and leisure, whereas the consumption tax affects only the second and third choices. But the fact that the consumption tax affects only two margins whereas the income tax affects three does not establish that the sum of the deadweight loss is less. I don't think we have to plead complete ignorance here; it is a matter of considering empirically the elasticities that are involved.

Bailey: Is it better to have a consumption tax than an income tax in relation to the labor-leisure choice that Dick Musgrave also spoke about? It depends on whether saving and work (or equivalently stated, whether consumption and leisure) are complements or substitutes. If they are complements and had fixed proportions like tea and sugar, or tea and milk, and if there was only one way to drink it, then there wouldn't be a second-best problem. The tax on consumption could be set, taking into account the labor-leisure consequence, which would occur in fixed proportions, and the result could be Pareto optimal. If they are neither complements nor substitutes, but right on the boundary line between those two cases, then it would be an improvement to go to a consumption tax because we would be harming only one margin instead of two, as Dick Musgrave put it. If consumption and leisure are substitutes, then it would be a bad idea to go to a consumption tax; it would make matters worse, like taxing butter when margarine isn't taxed, or the reverse. Being better on the saving-consumption margin would be at least partly or perhaps wholly offset by the greater harm it did on the labor-leisure margin. If consumption and leisure are complements, a change from an income tax to a consumption tax improves both margins.

Eisner: In shifting from an income-based tax to a consumption-based tax, we have to take care to define consumption. For instance, go back to Keynes's discussion of the motivations for saving and consumption in *The General Theory*. People save for precautionary motives, out of avarice, for the spirit of enterprise, and in order to have an opportunity to take advantage of what may develop. Two people with the same consumption but different wealth are in quite different positions. If consumption is fully measured appropriately, including certain implicit services that might be imputed from the ownership of wealth, an equalization of consumption could then be projected. But since that is not how consumption is measured, we are in rather treacherous territory in merely working toward equalization of consumption.

Graetz: If, in moving from an income tax to a consumption tax, we exclude savings from the base, it appears that higher rates will exist on what's left. If higher rates exist on what's left, and consumption is what's left, then both as a matter of efficiency and equity it is somewhat more important—depending on the precise rates—to tax consumption than it is under the current system. I don't think there is any doubt about that.

When we talk about a consumption tax, it is important to know whether things like imputed consumption (the farmer who is growing his own food) and consumption concealed in business transactions are in the base or not. There is also the question of including or excluding charitable contributions. I don't think you can talk about equity or efficiency considerations without describing exactly what is and what is not included in the base.

Bailey: Present exclusions from the income tax base such as food produced and consumed on the farm and perks for executives would be included in an ideal tax base whether we tax income or consumption. One should not make the mistake of assuming that it becomes more important to include them when we narrow the base of the tax from income to consumption. The incentive effects and tax consequences of exclusion or lax enforcement are the same for both tax-base concepts.

Graetz: Going from an income tax to a consumption tax strengthens the case for a significant tax on gifts and bequests, either through the consumption tax, through an accessions tax, or an estate tax, and taxing bequests is a method of redistribution that I prefer.

Buchanan: The appropriate tax base may be the present value of options. It is important to recognize, as Bob Eisner and others have talked about, that in one sense the possession of wealth does give one options. Options are something that haven't really been incorporated into our thinking on many of these things. The person with a tenured position really has options that many people without tenure don't have. If this sort of present value of option is followed as your base, the person with the tenured position would in fact have to pay a higher income tax or some kind of tax on the value of those options. It can be argued that the person with tenure somehow should have taken a lower wage in order to get the tenure.

But I don't see how that would change the need to tax the value of that option, because the person who chooses to work in Florida also takes a lower wage. So if he is taxed because of the added option of working where the sun shines, the other person must also be taxed on the tenure value. I'm confused here. I don't see how the market can take that into account. Perhaps it could be capitalized in land or something like that.

Musgrave: What about wealthy people who inherit so much wealth and earn so much income that they are not going to consume it, but leave it to the next generation? Consequently, we may get more and more concentration of wealth. Doesn't that worry you?

Feldstein: Well, I would point out that there are three things that can be done about that. The current system of estate taxes can be continued and even made more stringent by having much higher rates at the top if the accumulation of power in the form of wealth is really a strong concern; or we could have a wealth tax. There could also be a tax on even relatively low levels of wealth because of the kind of concern that Bob Eisner raised. Or there could be a tax with very low rates on a very large amount of wealth, a tax that focuses only on very substantial accumulations of wealth.

Let me now distinguish the wealth tax from an income tax. There is no particular reason why the tax rate that's appropriate for wealth should be essentially a tax on the interest income earned on that wealth. That is, the appropriate tax rate on wealth is not the income tax rate multiplied by the rate of interest, and that is what we have today.

Brennan: I am surprised that more attention hasn't been given to one point that seems to me to be obvious in favor of consumption taxation. This is the asymmetric way in which inflation affects labor and capital income. It is almost axiomatic that in a regime of high rates of inflation, almost independently of whether interest rates happen to be sticky, a tax based on nominal interest receipts is grotesquely inequitable. It seems likely that the future's rates of inflation are likely to be reasonably high for some time; therefore a consumption base is preferable to an income base quite independently of the conceptual issues that we have so far been talking about.

The purely conceptual case for an income tax hasn't yet been articulated sufficiently well. While I am in favor, on balance, of a consumption tax, that case should not be allowed to be made too easily. As I understand it, two social purposes might be served by the so-called double taxation of savings embodied in the income tax. The first is that future consumption and leisure are relatively complementary. In other words, the "optimal" tax might require us to tax future consumption more heavily than present consumption. Now I raise that as a possibility because it seems to me to be a point that's implicit in a lot of arguments about income as the appropriate tax base. The second is that if we agree that neutrality requires an income effect and no substitution effect, then obviously how we define a pure income effect depends crucially on the arguments in the utility function that one is dealing with. To argue that the only motive for savings is future consumption is not persuasive (Bob Eisner has already made this point). Accumulation for its own sake may also represent a motive for saving. The relationship that psychic rate of return bears to the money rate of return from such accumulation is not clear. But that doesn't compel me to argue that this psychic rate of return is zero.

For example, the rationale for a wealth tax over some range seems to depend crucially on the belief that assets enter into individual utility functions in a way that is independent of the future consumption returns that those assets generate. And once it is accepted that this is conceptually possible—and I think it has definite validity—then it becomes a very difficult empirical, introspective question whether an income tax base or a consumption tax base happens to be the more efficient or equitable.

Musgrave: When it comes to how wealth should be treated under a consumption tax, I find it crucial to include bequests as consumption. But assuming a world in which there are no bequests, I doubt whether an additional wealth tax is needed.

It's important to distinguish the motivations for which people hoard wealth from what one wants to tax. For instance, if the social power that results from wealth is a concern, one probably would not want to consider net worth, but gross worth. One would want to include the leverage that is surely a part of power. If one thinks simply of the pleasure of counting one's bills at night—the sort of sheer miser's satisfaction—then one considers net worth. Assume that the psychology of the miser is such that he does not count his money every night during the ten years, but he does all his counting just before he spends it. He counts it and then he spends it, and that's when he gets his miser's satisfaction. In that case, since there is a fixed relationship between the miser's satisfaction and the consumption satisfaction in the tenth year, and since consumption is taxed anyhow, it's irrelevant whether miser's satisfaction is included in the base. In that case an extra tax is not needed. But if the miser counts the money not only at the last minute, but also every night during these ten years, then the time-incidence of the miser's satisfaction is not postponed as in the first case. Since a consumption tax ought to be imposed at the time at which the consumption satisfaction occurs, the tax ought then to be divided into two parts, one which is collected annually on the miser's satisfaction and the other which is collected later at the time of consumption. So the timing factor adds a problem.

Feldstein: I like Dick Musgrave's analysis of the two kinds of returns to wealth, but I wonder if the market doesn't already take into account the return to miserliness in the rate of interest that it provides, so that people are in fact already paying, as in Martin Bailey's calculations of a couple of years ago, for the special privileges of enjoying untaxed miserliness in the lower gross rates of return.

The current tax law has many features that allow people with very high incomes to consume more than the amount on which they pay income taxes. Some of those features are loopholes that clever people could figure out how to close, and some are loopholes that don't get closed because they serve useful social purposes. It is advantageous to have people trading assets rather than being locked in, so capital gains are not taxed in full the same way income is taxed and so on. The consumption tax is a fair tax because it will tax the wealthy on their high consumption regardless of how they finance it, be it capital gains, state and local interest borrowing, or bequests. There are two separate issues: one is the progressivity and the other is the tax base. So we're talking about a progressive tax that will have the same effective rate, perhaps by income class, but will discriminate in favor of those who save more and against those who consume more.

Sunley: Treasury Department studies released in the last days of the Ford administration defined a comprehensive consumption tax base and hence determined the rates needed to get the same average tax burden by consumption classes. Their top marginal rate was 40 percent. One of the reasons the top marginal rate can be as low as 40 percent is that imputed income from home ownership is included in the tax base.

Feldstein: Do you think there would still be a distortion if charitable contributions were called consumption?

Sunley: I would call charitable contributions consumption.

Feldstein: Whether charitable contributions are called consumption is not my question. I want to ask whether you would still want a wealth tax if we did call charitable contributions consumption. Presumably the reason that one gets appointed to the board of directors of the symphony is because the board of directors expects a contribution from you. One of the things that you buy with that contribution is a sense of inner satisfaction, and another thing you buy is a seat on the board of directors—just as when you buy an automobile, you get a combination of satisfaction, status, and a variety of other things. Would it be different, then, from any other kind of consumption good? Would we have to tax wealth if we taxed the spending of wealth on certain kinds of activities that we now either don't tax at all because we treat it as a gift or actually tax in a negative way because we consider it a charitable contribution? Would we eliminate the whole problem of taxing wealth if we treated political gifts and gifts to charities as a form of consumption?

Mieszkowski: In the words of the old story, I can convince the poor to accept it; I'll leave it to you to tell me how the rich are going to give it up. It's essentially a political problem, one that we've ignored in much of our discussion. I would be intrigued to learn how one is to convince charities that instead of having a positive incentive for charitable giving, as we now do in the income tax, we're going to have a negative incentive because we're going to tax charitable giving as part of the consumption tax.

Wolfman: I have a short comment on Marty Feldstein's question. It seems to me that taxing the evidence of current power, such as that expressed in the charitable contribution, confronts only a part of the problem to which we allude when people suggest that wealth in itself is a factor to be considered apart from the income or the consumption that it fosters in a given period. I think the reason is that people tend to be deferential to the wealthy, and that tendency gives power to the wealthy. That's part of what we're talking about—various kinds of power, status, prestige. In a related sense, there is power because wealth gives an assurance of an income or a consumption stream that current income or current consumption alone does not. The mere taxation now of that borderline item as consumption does not affect the power that remains through the assurance of the continuous income stream.

Feldstein: The only people then with real wealth in a world of uncertain returns and high inflation are the people with tenure or civil service status.

Tolley: It seems to me that the arguments favor a consumption tax. Our treatment of private pensions indicates a progression in that direction. Now IRAs and Keough plans have been started. The direction that has proven to be feasible and politically realistic should be pursued to extend that treatment to defer taxes on savings on retirement a great deal further. That would be the first part of it. The second part of it would be to get at the loopholes and to say that gains in the tax system can be realized that probably have some hope of being politically acceptable if the tax loopholes are closed at the same time that the corporate income tax is reduced. In that way zero revenue losses would occur and compromises could be sought in the Ways and Means Committee. That's a possible immediate agenda.

Tax Rates

Surrey: I understand the base of the consumption tax. But how progressive should the rate structure be?

Ture: There is no ideal rate system, but fundamentally the idea is that it ought to be a flat rate, because it is not aimed or intended as a redistributive mechanism. It should be imposed at a rate that burdens equally consumption and saving uses of income, as well as consumption and capital formation uses of production capability. The addendum was to suppose the new tax can be chosen as part of a redistributive device. What I have urged, then, is that it ought to be used as a means of subsidizing the saving and capital accumulation of the poor. I would suggest doing it by creating a tax credit in lieu of a deduction for current saving.

If you want to use the tax mechanism effectively for purposes of income equalization, or for a move in that direction, the vehicle for doing so is to attempt to equalize, or more nearly equalize, the distribution of the ownership of capital. At some level of affluence and above, the saving-consumption choice should be treated neutrally; and at levels below that arbitrarily selected standard, subsidization of the saving would be what is wanted.

Graetz: If you are talking about moving today to a consumption tax as a policy prescription, and at the same time your're talking about moving to a flat-rate tax, then you would be very nonneutral toward redistributing wealth from those who have less today to those who have more, or redistributing consumption and income from those who have less to those who have more. If you are neutral, then you ought to keep the distribution constant. This is a problem of transitions and windfalls in changing from one system to another, and it is a very complicated problem. But if you try to achieve through a consumption tax base a distribution that is identical to that which we now have under the income tax, I suspect you are talking about a very steeply progressive rate schedule, probably in excess of 100 percent in some cases.

Cooper: I assume if you have a flat-rate consumption tax, then you are treating deferred consumption and present consumption equally. Then to subsidize saving by the poor, unless I missed something, is inefficient because you've created, at least for one class of people, a bias in favor of deferred consumption. Once you have a consumption tax, wouldn't the efficient way to redistribute wealth simply be to do it by imposing a wealth tax for upper income people and using it to create savings accounts for poor people?

Ture: I would rather argue that we should begin with a neutral tax treatment. If at that point you deem it appropriate to set income redistribution without loss of aggregate output as a high priority policy objective, that would most clearly call for providing a subsidy for the capital accumulation of the poor. To put the cards on the table, I would opt for the tax where I could say the relative price effects are minimal. A head tax would alter the relative price of being alive or not, but I suppose one could fudge that by saying the immediacy of response to that is likely to be relatively small, unlike virtually any other tax that one can dream up.

Feldstein: Was it the implication of your last remark that the income tax currently operates progressively in an empirical fashion? The evidence in Pechman's book seems to indicate the contrary, despite its nominal progression.

Buchanan: The Pechman-Okner results—that the major payments by income class indicate tax burden proportion—tell us nothing about the burden of the tax structure. The burden of the tax structure is more nearly measured by the nominal rate structure because to get to that proportion people have to accept all sorts of nonefficient uses of their income.

Brownlee: A great deal of consumption inequality that we observe is inequality by choice; some people choose not to work and would choose not to work even if our tax system didn't introduce distortion. Some people choose not to work because we do have distortion in the tax system. This point was implicit in Dick's discussion about potential consumption as a tax base. These differences in preferences should be taken into consideration.

Meiselman: I presume he meant some consumption of goods that had some pecuniary value.

Brownlee: In view of the possibility of inequality by choice, it's not at all obvious that changing the tax system will make things more equal or inequal. Nor is it obvious in which direction we ought to move if we agree that equality itself is good.

Halperin: I want to raise essentially the same question but from a different angle. People have suggested that a consumption tax can be developed that is just as progressive with respect to income as our current income tax. This, I take it, means we can have just as much redistribution across income classes as we now have. Does that suggest that efforts be abandoned to get more redistribution than we now have, or is the push in the other direction, for more redistribution? In other words, would people be willing to have a consumption tax that is just as progressive with respect to income as the nominal income tax?

Would the political or economic arguments against increased progression be reduced if we moved to a consumption tax? Or are we saying, "Let's take the progression we now have; that's as much as we're going to get"? However, the progression can be kept under a consumption tax without the bias, if there is one, against savings that results from an income tax.

Graetz: The optimal tax literature, in which a number of people around the table have done studies, suggests that some progressive rate is more nearly optimal than a proportional rate, although those studies depend on certain assumptions about social welfare functions, uniformity of preferences, and other assumptions that are highly questionable.

Buchanan: The flat-rate proportional tax would at least guarantee that the size of the government would go up roughly proportionally with national income. The progressive tax will, of course, guarantee that the government gets a big share of the pie automatically, given the political structure. It seems to me that there is a social problem that far transcends in importance all these redistribution questions and all these capital formation questions. Can we possibly get the size of the public sector under control before it eats us up? That is the problem, and it becomes very clear that a shift to a proportional tax would guarantee that the public sector would only increase proportionately with income.

Bailey: That could be done also by indexing the rate structure to average per capita income and keep the progressive tax structure.

Conclusion

Klein: If one likes playing chess and games like that, it's fun to listen to this discussion. But I am not sure how far we can usefully go with anything other than issues much more narrow and immediate than those we have been talking about. One thing that does seem clear to me about income distribution is that we have taken enormous strides in the past 10 years in distributing income toward the bottom of the income spectrum; in my political judgment that's all to the good. Further change must be incremental. I can't get very excited about the rather cosmic questions we have been raising. I would rather hear a discussion of questions related to moves in the direction of a consumption tax of the sort suggested by Bill Andrews.

Buchanan: It is just like playing chess. We do enjoy it. Or, as Armen Alchian would say, we play music. It's just mental exercise for the most part. Most of this discussion has really taken place within a complete institutional vacuum, with the assumption that some benevolent dictator is going to put down what we agree should be an optimal tax. Very little, if any, recognition of the political reality that exists has been acknowledged. If we look at political reality, the tax system is not going to be determined by some benevolent being or by some collection or committee that is going to listen to economists or tax lawyers. We should be taking a much harder look at the way the governmental structure in this country will in fact produce tax laws and tax structures.

Ture: If those who are faced with starvation are really of concern to us, we should recognize that there is a relatively small number of individuals or households whose economic situation is desperate. Then I would say we surely don't have to be concerned about any changes in the tax structure of the sort we've been talking about today. It's a matter of utter indifference what kind of taxes we levy. Dealing with that problem would be peanuts, fiscally. But if that is not the case, and the case really is whether or not we should equalize consumption, then I'm right back where I started from in the early part of my paper. I don't see anything that represents an ethical good in that.

Bibliography

Allen, R.G.D. *Macro-Economic Theory.* New York: St. Martin's, 1967.

Andrews, William D. "A Consumption-Type or Cash Flow Personal Income Tax." *Harvard Law Review* 87 (1974):1113-1188.

⸺. "Fairness and the Personal Income Tax: A Reply to Professor Warren." *Harvard Law Review* 88 (1975):947-952.

Berndt, Ernst R. "Reconciling Alternative Estimates of the Elasticity of Substitution." *Review of Economics and Statistics* 58 (1976):59-68.

Bittker, Boris L. "Accounting for Federal 'Tax Subsidies' in the National Budget." *National Tax Journal* 22 (1969):244-261.

Blum, Walter, and Kalven, Harry, Jr. *The Anatomy of Justice in Taxation.* Occasional Papers. Chicago: University of Chicago Law School, 1973.

⸺. *The Uneasy Case for Progressive Taxation.* Chicago: University of Chicago Press, 1963.

Boskin, Michael J. "Taxation, Saving and the Rate of Interest." Working Paper No. 135. New York: National Bureau of Economic Research, 1976.

Bradford, David F., and Rosen, Harvey S. "The Optimal Taxation of Commodities and Income." *American Economic Review, Papers and Proceedings* 66 (1976):94-101.

Buchanan, James M. "A Hobbesian Interpretation of the Rawlsian Difference Principle." *Kyklos* 29 (1976):5-25.

Budd, Edward C. "Postwar Changes in the Size Distribution of Income in the U.S." *American Economic Review, Papers and Proceedings* 60 (1970):247-260.

Cobb, C.W., and Douglas, Paul H. "A Theory of Production." *American Economic Review, Supplement* 18 (1928):139-165.

Daniels, Norman, ed. *Reading Rawls: Critical Studies on Rawls' A Theory of Justice.* New York: Basic Books, 1975.

Diamond, Peter A. "Incidence of an Interest Income Tax." *Journal of Economic Theory* 2 (1970):211-224.

Dworkin, Ronald. "The Original Position." In *Reading Rawls: Critical Studies on Rawls' A Theory of Justice,* edited by Norman Daniels, pp. 48-52. New York: Basic Books, 1975.

Eisenstein, Louis. *The Ideologies of Taxation.* New York: Ronald Press, 1961.

Feldstein, Martin. "On the Optimal Progressivity of the Income Tax." *Journal of Public Economics* 2 (1973):357-376.

⸺. "Incidence of a Capital Income Tax in a Growing Economy with Variable Savings Rates." *Review of Economic Studies* 41 (1974):505-513.

⸺. "On the Theory of Tax Reform." *Journal of Public Economics* 6 (1976):77-104.

⸺. "Social Security, Induced Retirement, and Aggregate Capital Accumulation." *Journal of Political Economy* 82 (1974):905-926.

Ferguson, Charles E. *The Neoclassical Theory of Production and Distribution.* London: Cambridge University Press, 1969.

Fiekowsky, Seymour. "The Impact of Taxation on Mineral Capital: The Case of Oil and Gas." In *Economics of the Mineral Industries,* edited by William A. Vogely, pp. 673-682. New York: American Institute of Mining, Metallurgical and Petroleum Engineers, 3rd ed., 1976.

Galvin, Charles O., and Bittker, Boris. *The Income Tax: How Progressive Should It Be?* Washington, D.C.: American Enterprise Institute, 1969.

Goode, Richard B. *The Individual Income Tax.* rev. ed. Washington, D.C.: The Brookings Institution, 1976.

Graetz, Michael J. "Assessing the Distributional Effects of Income Tax Revision: Some Lessons from Incidence Analysis." *Journal of Legal Studies* 4 (1975):351-368.

Hickman, Frederick. "Pension and Profit-Sharing Plans: The Quintessential Tax Shelter?" Department of Treasury Press Release, No. S-336, December 5, 1973.

Klein, William A. *Policy Analysis of the Federal Income Tax: Text and Readings.* Mineola, N.Y.: Foundation Press, 1976.

Musgrave, Richard A. *The Theory of Public Finance: A Study in Public Economy.* New York: McGraw-Hill, 1959.

_____. "ET, OT and SBT." *Journal of Public Economics* 6 (1976):3-16.

_____. "Growth with Equity." *American Economic Review* 53 (1963): 323-333.

_____. "Maximin, Uncertainty, and the Leisure Trade-off." *Quarterly Journal of Economics* 88 (1974):625-632.

Nozick, Robert. *Anarachy, State, and Utopia.* New York: Basic Books, 1974.

Okun, Arthur M. *Equality and Efficiency, the Big Tradeoff.* Washington, D.C.: The Brookings Institution, 1975.

Rawls, John. *A Theory of Justice.* Cambridge, Mass.: Harvard University Press, 1971.

Sandmo, Agnar. "Optimal Taxation: An Introduction to the Literature." *Journal of Public Economics* 6 (1976):37-54.

Scanlon, T.M. "Rawls' Theory of Justice." In *Reading Rawls: Critical Studies on Rawls' A Theory of Justice,* edited by Norman Daniels, pp. 171-179. New York: Basic Books, 1975.

Simons, Henry C. *Personal Income Taxation.* Chicago: University of Chicago Press, 1938.

Ture, Norman B., and Fields, Barbara A. *The Future of Private Pension Plans.* Washington, D.C.: The American Enterprise Institute, 1976.

United States Department of the Treasury. *Blueprints for Basic Tax Reform.* Washington, D.C.: United States Government Printing Office, January 17, 1977.

Warren, Alvin C., Jr. "Fairness and a Consumption-Type or Cash Flow Personal Income Tax." *Harvard Law Review* 88 (1975):931-958.

About the Contributors

Norman B. Ture is president of Norman B. Ture, Inc., an economic consulting firm, and is an adjunct scholar at the American Enterprise Institute for Public Policy Research. Dr. Ture received the M.A. and Ph.D. degrees in economics from the University of Chicago. He has been an analyst in the Tax Division of the Treasury Department and a staff member of the Joint Economic Committee developing programs for special studies in tax and expenditure policies. Dr. Ture was director of Tax Studies at the National Bureau of Economic Research for some seven years. He is the author of numerous books and articles specializing in the tax field.

Boris I. Bittker is Sterling Professor of Law at Yale. He is a specialist in federal taxation and corporate and constitutional law. He received the B.A. degree from Cornell and the LLB from Yale Law School. He was law clerk to the late Judge Jerome N. Frank of the U.S. Second Circuit Court of Appeals and was in the office of the General Counsel for the Lend-Lease administration in Washington before entering military service during the second world war. A member of the Yale Law faculty since 1946, Professor Bittker is the author of numerous books and articles on taxation.

Michael J. Graetz is professor of law at the University of Southern California. Before moving to Southern California, Professor Graetz was a member of the law faculty at the University of Virginia. He was educated at Emory University and the University of Virginia, receiving the LL.B. degree from Virginia in 1969. He played an important role in the formulation of recent tax policy, including the tax reform act of 1969 and its supporting regulations. He has published extensively in the field of income taxation.

Martin S. Feldstein is professor of economics at Harvard University. Professor Feldstein received the B.A. in economics from Harvard College and his graduate degrees, including the D.Phil., from Oxford University. He has served as a member of the economics panel of the National Science Foundation, a member of the Institute of Medicine of the National Academy of Sciences, and senior advisor and panel member of the Brookings Institution panel on economic activity. He has won the John Bates Clark Medal of the American Economic Association, an award given to the outstanding scholar in the American Economic Association under the age of 40. He is president of the National Bureau of Economic Research and has published extensive research in the areas of social insurance and capital formation.

Richard A. Musgrave is H.H. Burbank Professor of Political Economy at Harvard University. He received the M.A. and Ph.D. in economics from Harvard University. He has been a member of the board of editors of the *American Economic Review,* a member of the Executive Committee and vice president of the American Economic Association. He has served as an economic consultant to many governments, including our own and the governments of Columbia, Taiwan, Korea, Chile, and Japan. He has also been a consultant for the OECD and the World Bank. He was editor of the *Quarterly Journal of Economics* for seven years. He is one of the most distinguished figures in public finance and has written many books and articles on that subject.

About the Editor

Arleen A. Leibowitz is a member of the research staff of the Rand Corporation. She received the B.A. in 1964 from Smith College, the M.A. in 1965 from Columbia University (economics) and the Ph.D. in 1972 also from Columbia University (economics). Dr. Leibowitz was a research assistant (1971-1974) at the National Bureau of Economic Research, and a visiting assistant professor of economics at Brown University (1972-1975). She was Adjunct Professor at the University of Miami School of Law and a member of the research staff of the Law and Economics Center in the School of Law (1976-1977). She is a member of the American Economic Association and has contributed to the *American Economic Review,* the *Journal of Political Economy,* and the *Journal of Human Resources.*

LIBRARY OF DAVIDSON COLLEGE